Applied Data Analytics – Principles and Applications

RIVER PUBLISHERS SERIES IN SIGNAL, IMAGE AND SPEECH PROCESSING

Series Editors:

MONCEF GABBOUJ
Tampere University of Technology
Finland

THANOS STOURAITIS
University of Patras, Greece
and
Khalifa University, UAE

Indexing: All books published in this series are submitted to the Web of Science Book Citation Index (BkCI), to SCOPUS, to CrossRef and to Google Scholar for evaluation and indexing.

The "River Publishers Series in Signal, Image and Speech Processing" is a series of comprehensive academic and professional books which focus on all aspects of the theory and practice of signal processing. Books published in the series include research monographs, edited volumes, handbooks and textbooks. The books provide professionals, researchers, educators, and advanced students in the field with an invaluable insight into the latest research and developments.

Topics covered in the series include, but are by no means restricted to the following:

- Signal Processing Systems
- Digital Signal Processing
- Image Processing
- Signal Theory
- Stochastic Processes
- Detection and Estimation
- Pattern Recognition
- Optical Signal Processing
- Multi-dimensional Signal Processing
- Communication Signal Processing
- Biomedical Signal Processing
- Acoustic and Vibration Signal Processing
- Data Processing
- Remote Sensing
- Signal Processing Technology
- Speech Processing
- Radar Signal Processing

For a list of other books in this series, visit www.riverpublishers.com

Applied Data Analytics – Principles and Applications

Johnson I. Agbinya

Melbourne Institute of Technology
Australia

River Publishers

Published, sold and distributed by:
River Publishers
Alsbjergvej 10
9260 Gistrup
Denmark

www.riverpublishers.com

ISBN: 978-87-7022-096-5 (Hardback)
       978-87-7022-095-8 (Ebook)

©2020 River Publishers

This book is dedicated to two Agbinya descendants, Ireh Ihyeh Agbinya my last born (keep jumping), and Ihyeh Agbinya Agbinya the last born to Sunday Agbinya Agbinya my nephew. Both of you have brought joy to the family in difficult times.

Contents

Preface

Two decades ago, many electronic engineering and computing laboratories worldwide were engrossed in signal processing research. Signal processing experts normally heavily apply linear algebra and calculus to derive insight from signals. In recent times, signal processing concepts have been combined with statistical data analysis to create the new and exotic field of big data analytics. This reincarnation of signal processing has opened up data vaults held by many organisations as gold mines for high performing industrial data applications. Data analytics applies concepts derived from applied statistics, data mining, artificial intelligence and deep learning.

Many of the concepts in this book represent simplifications of the underlying foundations upon which deep learning and AI stand. In the book we have explained the underlying mathematical foundations and paid great attention to topics often found difficult by graduate and postgraduate students. The topics include Kalman filters, Markov chain, hidden Markov Modelling (HMM), neural networks, recurrent neural networks, convolutional neural networks, probabilistic neural networks, support vector machines, genetic algorithm, finite state machines and computational graphs. The chapters on statistics assume basic statistical foundations at the undergraduate level and thus offer deeper explanations on more difficult concepts including principal component analysis, statistical distributions by using probability-generating functions, moment generating functions, characteristic functions.

Most of the algorithmic foundations presented in the book are stable and have been in use for decades among signal processing and applied statistics experts. They are equally applicable to bioinformatics, data clustering and classification, data visualization, sensor applications and tracking.

The book is aimed at graduate studies and honours degree students will also benefit from its rich contents. It provides relevant mathematical tools and concepts required to capture, understand, analyse, design and develop data analytics frameworks. They will also simplify developing data analytic software programs and application of data analytics in various industries. By simplifying the algorithms and using relevant worked examples, the book lends itself to easy reading, following and lifelong learning of other concepts in use in data analysis.

A couple of chapters have inputs from my ex-postgraduate students and a third from a third source. These chapters results from long-term application of the concepts. We commend the book to instructors, research students and supervisors, course and algorithmic designers, developers in data and big data analytics and deep learning. By having functioning grasp of these foundations the reader will be in a strong position to interrogate and provide deeper practical insights into data analytics and deep learning in general.

Acknowledgement

We acknowledge the use of the rich resources in youtube. They have helped to guide simplification of difficult concepts through the excellent works of selfless teachers. I acknowledge my mentors who challenged me to demonstrate understanding of difficult concepts through easy to follow explanations.

List of Contributors

D. L. Dowe, *Monash University, Clayton, VIC, Australia*

Johnson I. Agbinya, *Melbourne Institute of Technology, Australia*

Rumana Islam, *University of Technology, Sydney, Australia*

Sid Ray, *Monash University, Clayton, VIC, Australia*

Tony Jan, *Melbourne Institute of Technology, Sydney, Australia*

Vidya Saikrishna, *Melbourne Institute of Technology, Melbourne Campus, Australia*

List of Figures

List of Tables

List of Abbreviations

1D	One Dimension
2D	Two Dimensions
3D	Three Dimensions
AFIS	Automatic Fingerprint Identification Systems
ANN	Artificial Neural Network
CNN	Convolutional Neural Network
DIMS	Digital Identity Management System
ELU	Exponential Linear Units
FSM	Finite State Machine
GA	Genetic Algorithm
GRBF	Gaussian Radial Basis Function
GRNN	Generalized Regression Neural Network
HMM	Hidden Markov Model
HPFSM	Hierarchical Probabilistic Finite State Machines
IoT	Internet of Things
KKT	Karush, Kuhn and Tucker
MGF	Moment Generating Function
MML	Minimum Message Length
MPNN	Modified Probabilistic Neural Network
NN	Neural Network
OM	Ordered Merging
PCA	Principal Component Analysis
PDF	Probabilistic Density Function
PFSM	Probabilistic Finite State Machines
PGF	Probability Generating Functions
PNN	Probabilistic Neural Network
PSVM	Parallel Support Vector Machine
PTA	Prefix Tree Acceptor
ReLU	Rectified Linear Unit
RNN	Recurrent Neural Network
RSSI	Received Signal Strength Indicator

1

Markov Chain and its Applications

1.1 Introduction

The current chapter is on a topic of great application in predicting outcomes of day-to-day processes based on observed probabilistic results from the past. This book chapter deals exclusively with discrete Markov chain. Markov chain represents a class of stochastic processes in which the future does not depend on the past but only on the present. The algorithm was first proposed by a Russian mathematician Andrei Markov. He was taught mathematics by another great mathematician Pafnuty Chebyshev at the University of St Petersburg. He was not particularly noted as a good student until he came under the mentorship of Chebyshev. Chebyshev was noted for his expertise in probability theory of which Markov chain is a part. Markov's first publication on Markov chain was in 1906. Since then the theory and applications of Markov chains has dramatically increased. In the recent past, like many other ancient mathematics theories, including Maxwell's equations, wavelets and a wide range of predictive mathematical algorithms, Markov chain has come to find its place in various practical applications. It has been applied in stock markets, weather prediction, spread of influenzas, susceptibility to breast cancer among women and various data analysis as we shall observe in this chapter. Markov chains model processes which evolve in steps which could be in terms of time, trials or sequence. Therefore, for example at each step, the process may exist in various countable states. When the process evolves, the system can remain in the same state or change (transition) to a different state during the time epoch. These movements between states are normally described in terms of transition probabilities. These transition probabilities allow us to predict into the future the possibility of the system being in a state, many time epochs later. We will get to see a few examples of this in this chapter.

The rest of this chapter introduces the concepts of Markov chains and defines the concept of states, state transition and how state transition diagrams are formed. It then adds the notion of state transition diagrams and how to use them in applications that require Markov chains.

1.2 Definitions

A stochastic process is a Markov chain if it possesses the Markovian properties. The Markovian property is simply that for the process the future and past states of the process are independent if its present state is known. To work with such concept, it is essential therefore to know the present state of the process.

Before we delve into the mathematics of Markov processes, suffice it to say that instead of burying the algorithm in a mass of variables and symbols as is often the case when Markov chains are discussed, the use of many variables and symbols will be simplified. Rather an explanatory approach with numerous examples will be given in discussing the subject.

The states X_n are discrete at time instance n. At time $n + 1$, the process depends only on the state it was at time n. For example, consider the spread of Ebola virus, X_n is the number of people that have Ebola virus at time n. The number of those who have the virus at time $n + 1$ is X_{n+1}. Following our definition of Markov chains, we therefore can write that the number of infected people at time $n + 1$ depends on those who were infected at time n (or $X_{n+1} \Rightarrow X_n$), where $\Rightarrow$ means 'depends only on'. The Markov process does not depend on $\{X_{n-1}, X_{n-2}, \ldots, X_0\}$.

1.2.1 State Space

The state space for the Markov chain is denoted by the letter S given by $S = \{1, 2, 3, \ldots, n\}$. In other words, the process can take n states. The state of the process is given by the value of X_n. For example, if $X_n = 4$ the process is in state 4. Therefore, the state of a Markov chain at time n is the value of X_n. The process cannot therefore be in two or more states at the same time.

Example 1: In most countries, some women are more susceptible to breast cancer than others. Female children who are born into families with history of breast cancer are more likely to develop breast cancer than women born into families without history of breast cancer. What are the possible states that a female child born into a society is susceptible to breast cancer?

Answer: *Historically, not all female children born in a society and into families with history of breast cancer eventually have breast cancer. Therefore, a* *female child in any society may either have a high risk of developing breast cancer or a low risk of developing breast cancer? This system has two states – High Risk and Low Risk.*

1.2.2 Trajectory

The trajectory or path of a Markov chain is the sequence of states in which the process has existed so far. We will denote the trajectory values as $s_0, s_1, s_2, \ldots, s_n$. In other words, the states took values as $X_0 = s_0, X_1 = s_1, X_2 = s_2, \ldots, X_n = s_n$.

1.2.2.1 Transition probability

From the above formulation, a Markov chain cannot be in two states at the same time. It can however change states from one state to another. When this happens, it is called a transition from state s_n to state s_{n+1}. This transition is represented with a probability value. A Markov chain can transit and remain in the same state from time n to time $n + 1$. Figure 1.1 is a state transition diagram for a two-state Markov chain. Initially at state 1, the system was at the High-Risk state. Assume that the probability that a woman at high risk of developing breast cancer today remains in the high-risk state a year later is 0.7. The probability that she transits to a low-risk state in a year from now is 0.3. This results in the state diagram given in Figure 1.1. The arrows show transition from an initial state to the next state one year from now.

Suppose the system was in the Low Risk as State 1. A similar transition diagram may be drawn and is given by Figure 1.2. Figure 1.3 shows the complete breast cancer process.

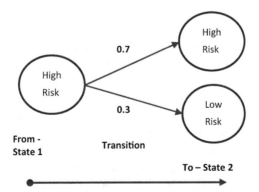

Figure 1.1 Example of state transition diagram starting at high risk.

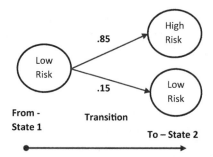

Figure 1.2   State transition from low risk as State 1.

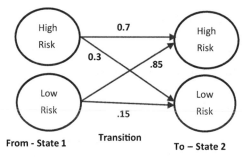

Figure 1.3   Combined-state transition diagram from State 1 to State 2.

Markov chains possess several properties. There are a few salient properties of state transition diagrams from Figures 1.1 and 1.2. First, the sum of transition probabilities from one state to any other state is always 1. The arrows in the transition diagram point from the initial state at time t to the state at time $t + 1$. Figures 1.1 and 1.2 do not show transitions back to the same state. This feature will be shown as we progress in the discussions. The combined state transitions from State 1 to State 2 is shown in Figure 1.3.

We have thus far demonstrated transitions from State 1 to State 2. Processes can also change state back from State 2 to State 1. This occurs at $t + 2$.

The combined state transition diagram is from only state 1 to state 2. Processes can change state back to the state they were in the previous time instance. That is like starting from a rainy day, transiting to a sunny day and on the third day having a rainy day which is a more practical and natural occurrence in nature. In Figure 1.4, we show a process with two states and transitions between state 1 and state 2. An example of such a process is when a woman that is susceptible to breast cancer changes state from being at no

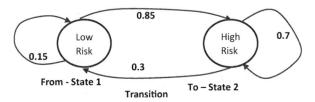

Figure 1.4 Two-state transition diagram.

risk of developing breast cancer, then became susceptible and high risk and next to low risk after being cured of breast cancer.

From Figure 1.4, the sum of transitions from each state is 1. From the state transition diagram, we observe the return to the same state.

1.2.2.2 State transition matrix

The state transition matrix is a matrix of all transitions. The rows are the starting points and the columns the end states. For breast cancer example in Figure 1.4, the state transition matrix is:

$$
\begin{array}{cc}
 & \text{LOW} \quad \text{HIGH} \\
\begin{array}{c} \text{LOW} \\ \text{HIGH} \end{array} &
\begin{bmatrix} 0.15 & 0.85 \\ 0.3 & 0.7 \end{bmatrix}
\end{array}
$$

The rows represent transitions from the same state and the columns represent the end state. For example, transition of the process from a low-risk state to a low risk has probability 0.15. Transition from low risk to high risk has probability 0.85.

So far, we have cases in which the transition probabilities are defined real fractions. Transition probabilities may also be calculated if all that is known is the relationships. Example 2 provides an illustration.

Example 2: *A university professor observes from students records that a first-year engineering student who fails his or her first-year mathematics subjects is three times as likely to also fail his or her core engineering subjects in the course. If the student passes the mathematics subjects, he/she is four times as likely to pass the core engineering subjects in the course.*

(a) Draw the probability tree and transition diagram for this problem.
(b) Write the transition matrix.

Solution: *The states for this problem are Fail and Pass. For the case when the student fails, the probability tree is given by the following diagram.*

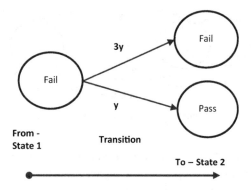

Since the sum of the probability transitions from a state is 1. Therefore, from the probability tree we can write the expression $3y + y = 1$ or $y = 0.25$. This results in the transition diagram

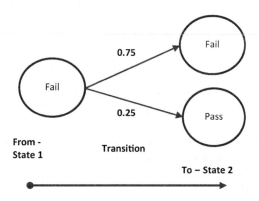

For the pass case, the probability tree diagram is

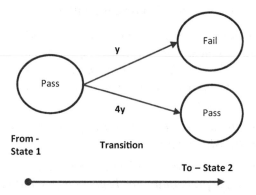

As in the Fail case, the probability equation is: $4y + y = 1$ or $y = 0.20$ and leads to the transition diagram

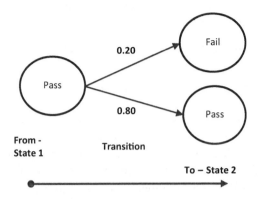

(a) The state transition matrix T is obtained by collecting the calculated transition probabilities and entering them into the matrix. The result is

$$T = \begin{array}{cc} & \overbrace{FAIL \quad PASS}^{} \\ \begin{array}{c} FAIL \\ PASS \end{array} & \begin{bmatrix} 0.75 & 0.25 \\ 0.2 & 0.8 \end{bmatrix} \end{array} \tag{1.1}$$

We can rewrite the calculations in a simple matrix framework as

$$T = \begin{bmatrix} 3y & y \\ y & 4y \end{bmatrix} \tag{1.2}$$

Solving these equations in this matrix row by row leads to the same results obtained previously. This leads to the same transition matrix

$$P = \begin{bmatrix} 0.75 & 0.25 \\ 0.2 & 0.8 \end{bmatrix} \tag{1.3}$$

When the information, are used, the transition probability is an internal information to a company from historical experience on the habits of its customers. The Initial State Vector could however represent the probabilities assigned publicly (for example as a market share) to the company. For example, consider two mobile phone operators who are competing for market share in a country, the initial state vector could be the publicly known market share. This value is available to most people outside the organisation as well as in the mobile phone companies. The state transition probabilities are however internal intelligence to each of the mobile phone company. Each company knows the number of its customers leaving to its competitor and the number sticking with them. The number sticking with them represents transition from the state to itself.

Problem 1: Draw the full-state transition diagram for Example 2.

1.3 Prediction Using Markov Chain

Forecasting of how a process evolves into the future is an extremely powerful weapon in marketing and promotion of services. It is also an essential element in how a viral infection spreads and how far the spread is likely going to be into the future. This section illustrates how to use Markov chain for forecasting or prediction. Whether it is weather prediction, rainfall prediction and electoral outcomes, Markov chain provides useful insights. Prediction with Markov chains requires knowledge of the probabilities of the initial state of the process and the state transition matrix. To illustrate this, we will use our example of engineering students with or without mathematics and how the student performs in core engineering subjects.

1.3.1 Initial State

Consider the initial state for a student who passed his or her first year mathematics. The initial state for the student is {Fail, Pass} = {0.2, 0.8}. Initially, the student's chance of failing core subjects is 0.2 and the chance of passing core subjects is 0.8. This probability tree is shown in Figure 1.5.

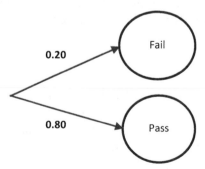

Figure 1.5   Initial-state probability tree.

Prediction of the future, which involves computing the probabilities that the student will pass his or her core units, is obtained by using the product of the initial-state probabilities and the state transition matrix. This formally is:

Prediction = Initial State × State Transition Matrix

Using our example, for the pass condition, this becomes

$$[0.2, 0.8] \begin{bmatrix} 0.75 & 0.25 \\ 0.2 & 0.8 \end{bmatrix} = [0.31, 0.69]$$

Problem Exercise

(1) Assume McDonalds (M) and Kentucky Fried Chicken (K) are the dominant fast food outlets in country Y. A local fast food outlet Jollof (J) wants to partner with either Maca (MacDonalds) or KFC (Kentucky Fried Chicken) and hires a market research company. The company was charged to find out which of the two fast food giants will have higher market share after one year. Maca currently has 60% and KFC 40% market share. The market research company found out the following probabilities.

Pr (M⇔M) is the probability of a customer remaining with Maca over one year = 0.75

Pr (M⇒K) is the probability of a customer switching from Maca to KFC over one year = 0.25

Pr (K⇔K) is the probability of a customer remaining with KFC over one year = 0.9

Pr (K⇒M) is the probability of a customer switching from KFC to Maca over one year = 0.1

The market research company is not sure yet what exactly to tell Jollof without making calculations based on transitions from company to company in the future.

(a) Draw the probability transition diagram

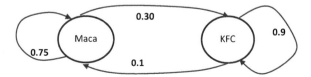

This is the current market share for Maca and KFC as a state transition diagram.

(b) What is the initial state probability vector (v)?

$$v = (0.6, \ 0.4)$$

(c) What is the market share for each of the two fast food companies after one year?

This problem is solved by using the time-step prediction model given in the previous section. The expression to use is:

$$Q_m = Q_{m-1} \cdot P = v \cdot P^m \tag{1.4}$$

where $m = 1$ and $Q_0 = v$

Therefore, the market share after one year is a matrix multiplication given by the expression:

$$P = \begin{matrix} & \overbrace{\begin{matrix} MACA & KFC \end{matrix}} \\ \begin{matrix} MACA \\ KFC \end{matrix} & \begin{bmatrix} 0.75 & 0.25 \\ 0.1 & 0.9 \end{bmatrix} \end{matrix} \tag{1.5}$$

The market share after one year is:

$$Q = v \times P^1 = (0.6, 0.4) \times \begin{bmatrix} 0.75 & 0.25 \\ 0.1 & 0.9 \end{bmatrix} = (0.49, 0.51) \tag{1.6}$$

(d) What professional advice should the market research company give to Jollof?

From the previous calculations, although McDonalds started with a higher market share of 60%, after one year, its market share has dropped to 49%, meaning that KFC has taken over with a market share of 51%.

(2) Predict the market share for McDonalds and Kentucky Fried Chicken into the future (Hint: this requires finding out that at some future time $t + m$ the value of the state vector probability remains unchanged even when m is increased).

From our calculations, it appears that the market share for McDonalds will continue to decline until when the number of clients who choose McDonalds and the number leaving are equal. At steady state, the following product holds

Initial state × Transition Matrix = Initial state

1.3.2 Long-run Probability

The long-run probability is also called the steady-state probability. It depicts when the initial state probability stops changing and remains constant. Let v be the initial state probability vector and P the state transition matrix, then the one-time step prediction Q is given by the expression

$$v_0 \cdot P = v_1 \tag{1.7a}$$

The subscript '1' on v shows that this is a one-time step prediction. Subsequent time step prediction re-uses the results of the previous calculation multiplied by the state transition matrix. That is, the two time-step prediction is

$$v_2 = v_1 \cdot P = (v \cdot P)P = v \cdot P^2 \tag{1.7b}$$

Similarly, the three time-steps prediction is

$$v_3 = v_2 \cdot P = v \cdot P^3 \tag{1.7c}$$

Therefore, the m-th time step prediction is therefore given by the expression

$$v_m = v_{m-1} \cdot P = v \cdot P^m \tag{1.7d}$$

This iterative process is used to predict future state probabilities of the process. At some value of m or time evolution, the values of $v_m = v_{m-1}$ after which multiplying with P does not lead to further improvements in the prediction. The long-run probability is often written as

$$v_\infty \cdot P = v_\infty \tag{1.8}$$

In other words, no amount of multiplication by the transition matrix leads to changes in the long-run probability vector.

1.3.2.1 Algebraic solution

In the above discussion, we have obtained the steady-state probability by repeated multiplication of the initial state vector with the state transition matrix and found that the initial state stabilises after some value of m. We might be able to solve for the steady-state vector without going through the numerous iterative multiplications to find m. Looking at the format of the long-run or steady-state solution, we derive some inferences.
Let

$$\begin{pmatrix} x & y \end{pmatrix} \times \begin{pmatrix} p_{11} & p_{12} \\ p_{21} & p_{22} \end{pmatrix} = \begin{pmatrix} x & y \end{pmatrix} \tag{1.9}$$

Assume that the values of the state transition probabilities are known, but the state vector $v = (x\ y)$ is unknown. Therefore, if the order of the vector and matrices are not too large, we can use simultaneous equations to find the values x and y. Occasionally, the initial-state vector may not be as simple as the one we have above but could contain many variables. When that is the case, matrix methods may be used.

1.3.2.2 Matrix method

In the matrix method, the solution for the steady-state vector is as follows. Let

$$vP = v \tag{1.10}$$

Subtract v from the left-hand side to obtain the equation

$$vP - v = 0$$
$$v(P - 1) = 0 \tag{1.11}$$

For a general matrix formulation, re-write this expression as

$$v(P - I) = 0 \tag{1.12}$$

where I is an identity matrix of the appropriate order. For example, the 2×2 identity matrix is $I = \begin{bmatrix} 1 & 0 \\ 0 & 1 \end{bmatrix}$. This allows the system of equations above to be solved. Several methods, including the eigenvalue analysis methods, may be used for solving for the state vector v.

Problem 4: Given the state transition matrix $\begin{bmatrix} 0.75 & 0.25 \\ 0.1 & 0.9 \end{bmatrix}$, find the steady-state vector v.

Solution:
Let

$$(x\ y) \times \begin{pmatrix} 0.75 & 0.25 \\ 0.1 & 0.9 \end{pmatrix} = (x\ y).$$

Then

$$(x\ y) \times \left(\begin{bmatrix} 0.75 & 0.25 \\ 0.1 & 0.9 \end{bmatrix} - \begin{bmatrix} 1 & 0 \\ 0 & 1 \end{bmatrix} \right) = 0$$

Solving this becomes

$$(x\ y) \times \left(\begin{bmatrix} -0.25 & 0.25 \\ 0.1 & -0.1 \end{bmatrix} \right) = 0$$

Two equations in unknown variables x and y result, which allows determination of the values x and y. The two equations are:

$$-0.25x + 0.1y = 0$$
$$0.25x - 0.1y = 0$$

Also $x + y = 1$. Hence, we can substitute for either of the variables in the above equations because $y = 1 - x$. Therefore, the first equation becomes

$$-0.25x + 0.1(1 - x) = 0$$
$$-0.35x = -0.1 \quad \text{and} \quad x = 0.2857 \quad \text{and} \quad y = 0.7143.$$

The steady-state vector therefore is $v = [0.2857 \ 0.7143]$.

1.4 Applications of Markov Chains

Markov chains have no memories, and as memoryless processes, the next state of the process depends only on the current state. Said differently, the current state depends only on the previous state.

One of the most useful data analyses with Markov chain is analysis of loan situations. Consider an "Unfair Bank". This bank has offered many housing loans, some under to customers who pay back their loans and regularly and others to loan defaulters. We will categorise Unfair Bank loan customers in four terms based on the type of loans the bank has offered to customers. The bank wants to understand why types of loans have a chance of being paid up and on time and the ones which are bad loans. Bad loans eat into the bank's revenue base and therefore need to be avoided as much as possible.

Paid Up Loans (PUL) are loans which have been fully repaid by customers.

Bad Loans (BL) are loans in which the customers have defaulted in paying.

Risky loans (RL) are loans for which the customers are categorised as medium to high risk. For such customers, defaulting in payment is to be expected.

Good loans (GL) are low-risk loans or customers. Such loans are alive and the customers are expected to pay back to Unfair Bank.

Therefore, the Unfair Bank model is a four-state model. The state transition probabilities are shown in Figure 1.6.

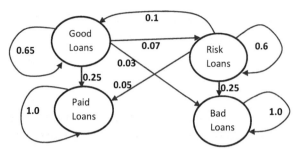

Figure 1.6   Unfair bank state transition diagram.

1.4.1 Absorbing Nodes in a Markov Chain

The state transition diagram contains two nodes where the transition probability at those nodes to themselves is 1.0. They show situations when if the process is in those nodes, there is no going out of them. Such nodes are defined as absorbing nodes. The two nodes are Bad Loans node and Paid Loans. Understandably, once a loan is completely paid, the state of the loan cannot change any further.

There is still need to define the initial state of the process. This is simpler than we imagine. From known statistics at the start of when the decision to offer loans, there can be no bad loans and paid up loans. Their probabilities are thus individually zero. From historical data of loans portfolio, it is possible to obtain the proportions of Good Loans to Risky loans. Let this be 0.65 and 0.35, respectively. Therefore, the initial state vector and state transition matrix for the Unfair Bank loan Markov process are, respectively:

$$v = \begin{matrix} Good & Risky & Bad & Paid \\ (0.650 & 0.350 & 0.00 & 0.00) \end{matrix} \tag{1.13}$$

The state transition matrix is

$$P = \begin{matrix} & \overbrace{\begin{matrix} Good & Risky & Bad & Paid \end{matrix}} \\ \begin{matrix} Good \\ Risky \\ Bad \\ Paid \end{matrix} & \begin{bmatrix} 0.65 & 0.07 & 0.03 & 0.25 \\ 0.1 & 0.6 & 0.25 & 0.05 \\ 0 & 0 & 1.0 & 0 \\ 0 & 0 & 0 & 1.0 \end{bmatrix} \end{matrix} \tag{1.14}$$

Problem 5: Compute the state of the bank loans at one year time step.

This is given by the expression

$$v_1 = v_0 P$$

$$= \begin{bmatrix} 0.650 & 0.350 & 0.00 & 0.00 \end{bmatrix} \times \begin{bmatrix} 0.65 & 0.07 & 0.03 & 0.25 \\ 0.1 & 0.6 & 0.25 & 0.05 \\ 0 & 0 & 1.0 & 0 \\ 0 & 0 & 0 & 1.0 \end{bmatrix}$$

$$= \begin{bmatrix} 0.4575 & 0.2555 & 0.107 & 0.18 \end{bmatrix}$$

At the end of the first year of loan, Good Loans = 0.4575, Risky Loans = 0.2555, Bad Loans = 0.107 and Paid Loans = 0.18. While the number of Risky loans is high, the number of paid loans is still small at the end of the first year.

Exercise: (a) Compute the steady-state probability vector for the Unfair Bank loan process. (b) What does the steady-state probability vector tell the Unfair Bank about its loans?

2

Hidden Markov Modelling (HMM)

2.1 HMM Notation

Markov chain processes constrain all processes to within one zone of existence in which all the states are observable. Hidden Markov modelling (HMM) differs from Markov chain in that the space within which a process expresses itself is divided into two disjoint spaces. The first space is similar with the space in Markov chain and contains all the hidden states. Outside this space exist all the expressions of the HMM process. These spaces are shown in Figure 2.1. Expression of the process in each space is described with a transition matrix and an emission matrix. Both matrices will be discussed in this chapter as we progress in the study of HMM.

In Figure 2.1, the arrows indicate emissions from the hidden states in the hidden states space. Hidden states result in the state transition matrix A to be described soon. Emissions are observable within the observable space and described by emission or confusion matrix B. This too will be described soon.

The HMM requires specification of the variables to be used in the analysis. Consider the set of N states given as $S = \{S_1, S_2, \ldots, S_N\}$. This set represents the observable states, visible to an observer there is a sequence of hidden states described by the sequence

$$Q = \{q_1, q_2, \ldots, q_T\} \tag{2.1}$$

The HMM process is described by the model:

$$\lambda = (A, B, \pi) \tag{2.2}$$

where A is the state transition probability matrix (for the hidden states). It is similar with the state transition matrix in first-order Markov chains. It is formed from state transition probabilities given by expression

$$a_{ij} = P(q_{t+1} = j | q_t = i) \tag{2.3}$$

17

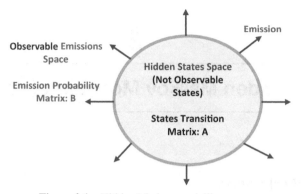

Figure 2.1   Hidden Markov modelling spaces.

Note that in all cases the transition probabilities are positive and sum to 1:

$$a_{ij} \geq 0, \ \forall i, j$$

$$\sum_{j=1}^{N} a_{ij} = 1, \ \forall i \tag{2.4}$$

This variable is the state transition probability at time $t + 1$ from state i to state j. The other set of variables include the observation probability distribution B. A typical example of an HMM is the stock market. The state of the stock market is S = {positive, neutral, negative}. This is called the mood of the market. The observables of the stock market is O = {increasing, decreasing} price.

2.2 Emission Probabilities

This section describes what happens outside the Hidden States Space (HSS) or in the Observable Emission Space (OES). For the discrete HMM, the probabilities of the observables are called 'emission probabilities'. The name arise from the fact that the hidden states are not observable themselves. It is only possible to observe what the hidden states emit (for example, a person wearing a winter coat suggests the weather is wintery, and the wearing of the winter coat is an emission or what is seen as a result of the weather condition), what the process spits out called measurements are specified as

$$b_j(k) = P(o_t = k | q_t = j); \quad 1 \leq k \leq M \tag{2.5}$$

This says that given that the system was in state j, $b_j(k)$ is the emission probability of observation k.

These observations or measurements or data are manifestations of hidden states. They are data that can be measured. While we are not able to see the hidden states, we are however capable of observing their effects. For example, assume you are a hermit and live in an enclosed house and have not ventured outside your house for many days. Each day you look outside your window. On the first day, you see a person with an umbrella, the second day the person is wearing shorts and a singlet, and on the third day the person is dressed in winter coats all covered up. From these three observations, one may guess that the first day was a rainy day, the second day was sunny, and the third day is wintery or a cold day. All that the hermit saw were the effects of rainy, sunny and wintery days.

This is the probability that the observation at time t is k given that the process was in state j. The elements $b_j(k)$ are taken from the so-called confusion or emission matrix. The values of the emission probabilities form elements of the so-called confusion matrix or emission matrix and represent the probability of a manifestation or observation. They are a consequence of having hidden states that emit the observations.

For the analog case, the emission probabilities are

$$b_j(x) = P(o_t = x | q_t = j); \quad 1 \le k \le M \tag{2.6}$$

This is the probability that you will see the symbol x (sunny day dressed woman, wintery day dressed woman or rainy day dressed woman) given that the system is in state j (sunny, rainy or wintery). For the rest of the book, we have limited discussion to the discrete case. The initial state distribution is similarly given as

$$\pi(j) = P(q_1 = j) \tag{2.7}$$

Figure 2.2 Emission or observations from the weather HMM.

For illustration consider what these women are wearing during three different days. The first an open shoulder clothing with sunglasses, the second a full winter coat with a head scarf and the third picture is of a woman using an umbrella. These three photos are the emissions from hidden states. They are observables on the days being considered, that is sunny, wintery and rainy days.

2.3 A Hidden Markov Model

This section describes how to model what is inside the Hidden States Space. We will use the weather to illustrate the Markov model. Assume the weather model is a first-order Markov process with state transition probabilities as shown in Figure 2.3. The probability of moving to the next state depends only on the current state.

To ease understanding of the concept, the transition probabilities are written in more explicit forms in this section. Starting from the Sunny state, the following are the transition probability expressions:

At Sunny State

$$a_{sw} = (q_w = S_w | q_s = S_s) = 0.1$$
$$a_{sr} = (q_r = S_r | q_s = S_s) = 0.25$$
$$a_{ss} = (q_s = S_s | q_s = S_s) = 0.65$$

Figure 2.3   First-order Markov model of hidden states.

At Wintery State

$$a_{wr} = (q_r = S_r | q_w = S_w) = 0.25$$
$$a_{ws} = (q_s = S_s | q_w = S_w) = 0.10$$
$$a_{ww} = (q_w = S_w | q_w = S_w) = 0.65$$

At Rainy State

$$a_{rs} = (q_s = S_s | q_r = S_r) = 0.40$$
$$a_{rw} = (q_w = S_w | q_r = S_r) = 0.30$$
$$a_{rr} = (q_r = S_r | q_r = S_r) = 0.30$$

2.3.1 Setting up HMM Model

To set up an HMM model, we need a set of states. We also need a set of transition probabilities and an initial-state distribution probability to start with in the process. This is shown in Figure 2.3. The process has a set of states, a set of state transition probabilities and initial state, which we describe as

$$S = \{S_1, S_2, \ldots, S_N\} = \{S_{sunny}, S_{rainy}, S_{wintery}\} \qquad (2.8a)$$
$$a_{ij} = P(q_{t+1} = S_j | q_t = S_i) \qquad (2.8b)$$
$$\pi(i) = P(q_1 = S_i). \qquad (2.8c)$$

This is the initial-state distribution. In our current example, we have three states, so N = 3. Note that generally $\sum_{j=1}^{N} \pi_j = 1$. From Figure 2.2, which provides the state transition diagram, the state transition matrix can be obtained and is:

$$A = \begin{pmatrix} 0.65 & 0.25 & 0.10 \\ 0.40 & 0.30 & 0.30 \\ 0.10 & 0.25 & 0.65 \end{pmatrix}$$

Since the process is Markovian, the state-transition probabilities from each state sums to 1. They are shown by the rows of the state transition matrix A. The initial state distribution for the three-state process is also

$$\pi = \begin{bmatrix} 0.65 & 0.25 & 0.10 \end{bmatrix}$$

Notice that the initial state distribution has three probabilities, which sum to 1 for sunny (0.65), rainy (0.25) and wintery (0.10).

Example 1: What is the probability of getting the series of the following images?

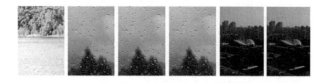

This series represent sunny, rainy, rainy, rainy, wintery, wintery

$$A = \begin{pmatrix} 0.8 & 0.15 & 0.05 \\ 0.38 & 0.6 & 0.02 \\ 0.75 & 0.05 & 0.2 \end{pmatrix}; \quad \prod = \begin{pmatrix} 0.7 & 0.25 & 0.05 \end{pmatrix}$$

Solution: The solution to finding the probability of the sequence can be found from the data given in the above picture. This is obtained from starting with the probability of the initial state that we are (0.7) and the product of the other transition probabilities: (sunny, rainy, rainy, rainy, snowy, snowy). The probability is:

$$p = p(S_{sunny}) \cdot p(S_{sunny}|S_{rainy}) \cdot p(S_{rainy}|S_{rainy})$$
$$\cdot p(S_{rainy}|S_{rainy}) \cdot p(S_{wintery}|S_{rainy}) \cdot p(S_{wintery}|S_{wintery})$$
$$= 0.7 \times 0.15 \times 0.6 \times 0.6 \times 0.02 \times 0.2$$
$$= 0.0001512$$

Why did we not start by using 0.8? That is because that probability represents the fact that we were in the sunny state today and the next day it remained sunny.

2.3.2 HMM in Pictorial Form

The observables are the types of clothes worn by the woman. We are not aware of the type of weather that led her wearing those clothe type. We would like to know what the weather is outside that has caused her to wear in each occasion those types of clothes. The types of clothes worn are the observables. The 'weather' state that leads to the observables are S and called hidden states. They emit observables. The emission probability that she wears a type of clothing is

$$b_j(k) = P(o_i = k|q_t = S_i) \tag{2.9}$$

b_j is the probability of the observation k given that the process is in the hidden state S_i.

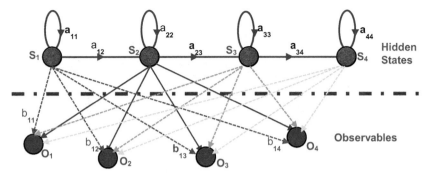

Figure 2.4   Illustration of hidden Markov model.

S_i is the current state from where the observation is emitted. The four-state hidden process when operational looks pictorially like the following diagram (Figure 2.4).

In general, from a hidden state i, the sum of emission probabilities from the state is equal to one. That is, for $i = 1$ to N, where N is the number of hidden states

$$\sum_{k=1}^{K} b_{ik} = 1 \qquad (2.10)$$

K is the number of observations. That is, for the process

$$b_{i1} + b_{i2} + b_{i3} + \cdots + b_{iK} = 1 \qquad (2.11)$$

For the three-state weather model example, in this section, we combine the HSS and OES models into a single Hidden Markov model. Figure 2.4 is the resulting HMM when the weather model is used as an example.

$$
A = \begin{array}{c} \\ Sunny \\ Wintery \\ Rainy \end{array}
\begin{array}{ccc} Sunny & Wintery & Rainy \end{array}
\begin{pmatrix} 0.65 & 0.25 & 0.10 \\ 0.40 & 0.30 & 0.30 \\ 0.10 & 0.25 & 0.65 \end{pmatrix}
$$

Notice that the sum of the emission probabilities from each hidden state to the observables equals 1. For clarity we expand the circle surrounding the hidden states in Figure 2.5 showing the state transition matrix for the hidden states.

For the hidden states, the transmission probabilities from one hidden state to another also sum to 1. In the hidden state, we often derive the

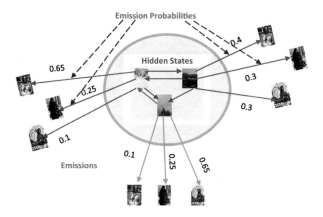

Figure 2.5   Hidden and observable states.

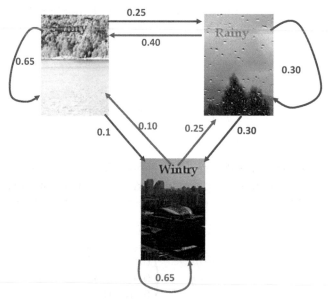

Figure 2.5a   Hidden states inside the circle of Figure 2.5.

transition matrix. In the observable states we also create the emission matrix (or also called the confusion matrix).

We will now use the above information in the pictures and ask the following question in the figure below.

Example: Suppose we have the following sequence of observations

What is the probability of this time series? Assume that you are given the transition probability, initial state and the emission probability shown.

$$A = \begin{pmatrix} 0.8 & 0.15 & 0.05 \\ 0.38 & 0.6 & 0.02 \\ 0.75 & 0.05 & 0.2 \end{pmatrix}; \quad \prod = \begin{pmatrix} 0.65 & 0.25 & 0.1 \end{pmatrix};$$

$$B = \begin{pmatrix} 0.6 & 0.3 & 0.1 \\ 0.25 & 0.6 & 0.15 \\ 0.0 & 0.5 & 0.5 \end{pmatrix}$$

Given the above observations, what is the probability of the sequence of observation shown?

The way to write this question mathematically is shown in the equation below

$$P(O) = P\left(O_{coat} \quad O_{coat} \quad O_{sunglass}, \quad O_{umbrealla}, \quad \cdots \quad O_{umbrealla}\right)$$
$$= \sum_{all\ Q} P(O|Q)P(Q)$$
$$= \sum_{(q_1,\dots,q_7)} P\left(O|q_1,\dots,q_7\right) P\left(q_1,\dots,q_7\right)$$

Over the seven days, various combinations of weather patterns are possible and need to be considered in arriving at the solution to the problem. The probability for this sequence is given by the expression

$$P(O) = P\left(O_{coat} \quad O_{coat} \quad O_{sunglass}, \quad O_{umbrealla}, \quad \cdots \quad O_{umbrealla}\right)$$
$$= (0.65 \times 0.8^6) \times (0.3^2 \times 0.1^4 \times 0.6) + \cdots$$

If it is all sunny days, we start with the initial probability 0.65 (sun) multiplied by 0.8 (six times or six sunny days). There is 0.1 probability of seeing an umbrella when it is sunny, 0.2 probability of seeing a coat when it is sunny and 0.3 probability of coat when it is sunny. The above situation is only for one possibility. We need to worry about all the possibilities on sunny days not just one.

2.4 The Three Great Problems in HMM

2.4.1 Notation

λ is the HMM model formed with three triplets
A is the transition probability matrix
B is the emission matrix
π is initial state distribution

2.4.1.1 Problem 1: Classification or the likelihood problem (find $p(O|\lambda)$)

Given an observation sequence $O = \{o_1, o_2, \ldots, o_T\}$ and the model $\lambda = (A, B, \pi)$, what is the probability of occurrence of the given sequence of observations? This is also termed a ***recognition or classification*** problem. It assumes that there exist HMM models. Which one is the one that likely generates the observation sequence?

2.4.1.2 Problems 2: Trajectory estimation problem

Given an observation sequence and the model $\lambda = (A, B, \pi)$, what state sequence is optimal in some sense which best explains the observation? This is also called the ***decoding*** problem. It seeks to find the optimal state sequence that produces the observation sequence $O = \{o_1, o_2, \ldots, o_T\}$.

2.4.1.3 Problem 3: System identification problem

How should the model parameters $\lambda = (A, B, \pi)$ be adjusted to maximize the likelihood? This is a system ***identification or learning*** problem and requires HMM to be trained to obtain the best model that describes the data. The problem here is to find λ to maximize the likelihood $p(O|\lambda)$. Sometimes it is also called the expectation maximization algorithm or EM HMM.

2.4.2 Solution to Problem 1: Estimation of Likelihood

A dirty approach that could be used to find the likelihood is to use a brute force. The goal is to find the probability $p(O|\lambda)$. The brute force approach requires that all possible state trajectories be interrogated or tried. For each trajectory, estimate the probability of the observed sequence or a likelihood value. This is called the exhaustive search procedure and is computationally expensive. Consequently, which ever likelihood value is highest, that is taken as the result. Unfortunately, with the number of states S, this computation problem is of order OS^N. This approach is not practical particularly when the process has so many states. We can use the time invariance of the problem to reduce the computational cost of the problem. There is therefore a better and quicker approach for solving the problem. It is called Forward–Backward procedure. It is found to be a lot quicker and easier to undertake.

2.4.2.1 Naïve solution

This section introduces the naïve solution. A brute force approach is used instead of a clever algorithm.

Normally, the observations are independent. We assume the observations are given by the sequence $Q = (q_1, q_2, \ldots, q_T)$

The probability of the T independent observations given the process is the product of probabilities and is written as in Equation (2.12)

$$P(O|q, \lambda) = \prod_{t=1}^{T} P(o_t|q, \lambda) = b_{q_1}(o_1)b_{q_2}(o_2)\ldots b_{q_T}(o_T) \qquad (2.12)$$

Since the probability of any state sequence is well known to be

$$P(q|\lambda) = \pi_{q1}a_{q1q2}a_{q2q3}\ldots a_{q(T-1)qT} \qquad (2.13)$$

Finally, the solution we are looking for is given by the expression

$$P(O|\lambda) = \sum_{q} P(O|q, \lambda) \cdot P(q|\lambda) \qquad (2.14)$$

This solution takes a very long time to conclude and is not efficient. There are N^T states paths and each of the path requires O(T) calculations. The overall cost of the naïve solution is therefore $O(N^T T)$. In the next section, a fast method of arriving at the same solution is given. It is called the forward recursion.

2.4.2.2 Forward recursion

In the forward recursion, the efficiency of the computation depends on the use of the Markovian concept. In the Markov process, the probability of transition depends on the state the process is now and not on the past and not how one arrives at the current state. In this section, the algorithm is derived.

(1) Define the forward recursion variable α. Given the model the forward recursion variable is

$$\alpha_t(i) = P(o_1, o_2, \ldots, o_t, q_t = i|\lambda) \qquad (2.15)$$

where $\alpha_t(i)$ is the probability of observing a partial sequence $(o_1, o_2, \ldots, o_t)$ when the model is given and the system is at state i at time t. The recursion begins with the following expression:

(i) Initialization
Let

$$\alpha_1(i) = \pi_1 b_i(o_1) \tag{2.16}$$

This is the probability of starting in state i, π_1 and seeing the first symbol when in state i, $b_i(o_1)$.

(ii) Induction
This depends on the probability that I land at state j from all possible N states and observing an emission j. This is given by the expression

$$\alpha_{t+1}(j) = \sum_{i=1}^{N} (\alpha_t(i)a_{ij})b_j(o_{t+1}) \tag{2.17}$$

(iii) Termination
The end of the recursion is the sum of the recursion variable for all time. This is given by the expression

$$P(O|\lambda) = \sum_{i=1}^{N} \alpha_T(i) \tag{2.18}$$

This is the probability of the whole sequence as a sum of observation sequence ending in state i for all i. The computational cost of the recursion is a lot smaller than that for the exhaustive search and is $O(N^2T)$.

The forward recursion may be viewed as a multi-channel Markov chain of N states. The initial state distribution is replaced for each channel with the product of the channel transition probability and the observation probability. The induction step is similar with computing the product $v_{t+1} = v_t P$ followed with the termination step. The forward recursion between two-time steps is shown in Figure 2.6.

2.4.2.3 Backward recursion

The backward recursion shown in Figure 2.7 starts from the time step $t + 1$ and works backward in time to time t.

For the backward recursion, we define the backward variable $\beta_t(i)$, where

$$\beta_t(i) = P(o_{t+1}, o_{t+2}, \ldots, o_T | q_t = i, \lambda) \tag{2.19}$$

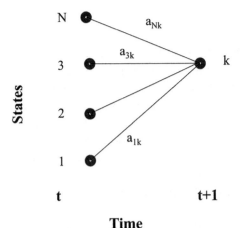

Figure 2.6 Forward recursion in HMM.

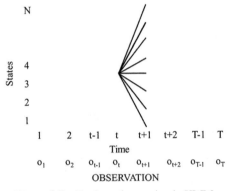

Figure 2.7 Backward recursion in HMM.

$\beta_t(i)$ is the probability of observing the partial sequence $(o_{t+1}, o_{t+2}, \ldots, o_T)$ at state $q_t = i$ given the Markov model, λ. The steps in the backward recursion are as follows:

(i) Initialization
Let
$$\beta_T(i) = 1$$

(ii) Induction
$$\beta_t(i) = \sum_{j=1}^{N} a_{ij} b_j(o_{t+1}) \beta_{t+1}(j); \quad 1 \leq i \leq N; \ t = T - 1, \ldots, 1$$

$$(2.20)$$

Example

2.4.2.4 Solution to Problem 2: Trajectory estimation problem

The objective of this problem is to find the path or state sequence $\{q_1, q_2, \ldots, q_T\}$, which maximizes the likelihood function $P(q_1, q_2, \ldots, q_T | O, \lambda)$. The solution is given as the Viterbi algorithm. The Viterbi algorithm is used not only in HMM applications like speech recognition but also in telecommunications in convolution encoders to detect symbols in noisy channels. The objective of this section is to present the algorithm in a simplified form so that it can be used more extensively. There is an extensive body of knowledge in the public domain for further interrogation on the topic if required.

Following the form given by Rabiner [1], the Viterbi algorithm is given in the following steps. The Viterbi algorithm is similar to the forward procedure. The difference is that instead of summation it uses maximization over previous states.

(1) Initialization

$$\delta_1(i) = \pi_i b_i(o_1), \ 1 \leq i \leq N \tag{2.21}$$
$$\psi_1(i) = 0$$

The zero indicates there were no previous states.

(2) Recursion

$$\delta_t(j) = \max_{1 \leq i \leq N} [\delta_{t-1}(i) a_{ij}] b_j(o_t) \tag{2.22a}$$
$$\psi_t(j) = \arg \max_{1 \leq i \leq N} [\delta_{t-1}(i) a_{ij}] \tag{2.22b}$$

$$2 \leq t \leq T, \quad 1 \leq j \leq N$$

(3) Termination

$$P_T^* = \arg \max_{1 \leq i \leq N} [\delta_T(i)] \tag{2.23}$$
$$q_T^* = \arg \max_{1 \leq i \leq N} [\delta_T(i)]$$

(4) Path Backtracking (retrieving of optimal path sequence)

$$q_t^* = \psi_{t+1}(q_{t+1}^*); \quad t = T - 1, \ T - 2, \ldots, 1 \tag{2.24}$$

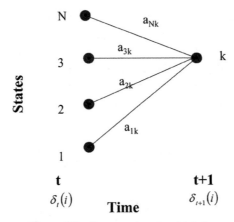

Figure 2.8 States at time t and $t + 1$.

Example: Detection of Symbols

This section gives and describes two Viterbi tables. The first table describes the state transitions. The second table gives what the output bits from the process should be when state transitions occur. These tables, for example, describes completely the behavior of a rate 1/2, $K = 3$ convolution encoder. The first table is described first.

2.5 State Transition Table

Consider the following table which provides information on which describes the state transition procedure. It gives the next state when the current state is given as well as the input and output bits.

2.5.1 Input Symbol Table

	Next State, if	
Current State	Input = 0:	Input = 1:
00	00	10
01	00	10
10	01	11
11	01	11

In the first row, if the current state is 00 and the input bit is '0', the next state is 00. If the input bit is '1', the next state is 10. For the second row, if the current state is 01 and the input bit is 0, the process changes state to 00 and if the input bit is a '1', the process changes state to 10. In the third row, if the current state is 10 and the input bit is '0', the state changes to state 01. If the input bit is a '1', the state changes to state 11. In the fourth row, if the current state is 11 and the input is '0', the process changes state to 01 and when the input bit is a '1', the next state becomes 11.

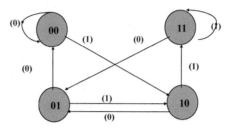

This state transition diagram also provides further explanation on the transitions given in the table above. Input bits are shown inside brackets. States are shown in green circles. Input bits are shown inside brackets. States are shown in green circles.

2.5.2 Output Symbol Table

This table provides the guide on what the output symbol should be at each state as a function of input bits to the process or system. From the output symbol table, in the first row, if the current state is 00 and the input bit is a '0', the output symbol from the process is 00. If the input bit is a '1', the output symbol becomes 11.

| | Output Symbols, if | |
Current State	Input = 0:	Input = 1:
00	00	11
01	11	00
10	10	01
11	01	10

From the second row, if the current state is 01 and the input bit is a '0', the output symbol from the process is 11. If the input bit is a '1', the output symbol becomes 00. From the third row, if the current state is 10, and the input bit is a '0', the output symbol from the process is 10. If the input bit is a '1', the output symbol becomes 01. In the fourth row, if the current state is

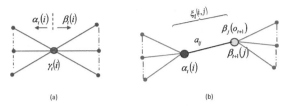

Figure 2.9   Trellis fragments (a) Occupation (b) Transition in calculation of likelihood.

11, and the input bit is a '0', the output symbol from the process is 01. If the input bit is a '1', the output symbol becomes 10.

2.6 Solution to Problem 3: Find the Optimal HMM

The objective of the problem is to compute an optimal HMM $\lambda = (A, B, \pi)$ to maximize the likelihood $P(O|\lambda)$. This problem is solved using an iterative process called the Baum–Welch algorithm, which we now discuss in reasonable details. The algorithm is also called expectation maximization (EM). Four processing steps are required.

2.6.1 The Algorithm

(1) Let the initial model be λ_0
(2) Compute a new HMM model λ based on λ_0 and observation O.
(3) If the difference $\log P(O|\lambda) - \log P(O|\lambda_0) < \Delta$, stop
(4) Else set $\lambda_0 \leftarrow \lambda$ and go to step 2 and repeat

To go through the algorithm, two variables will be defined as follows:

Define $\xi(i, j)$ the probability of being in state i at time t and being in state j at time $t + 1$. This probability is computed with the expression

$$\xi(i, j) = \frac{\alpha_t(i)a_{ij}b_j(o_{t+1})\beta_{t+1}(j)}{P(O|\lambda)} = \frac{\alpha_t(i)a_{ij}b_j(o_{t+1})\beta_{t+1}(j)}{\sum_{i=1}^{N}\sum_{j=1}^{N}\alpha_t(i)a_{ij}b_j(o_{t+1})\beta_{t+1}(j)}$$
(2.25)

This probability procedure is described with the following trellis diagram
    The objective of this section is to re-estimate the model as $\lambda_i = (A_i, B_i, \pi_i)$. To achieve this, the initial state probabilities are estimated as

$$\hat{\pi}_i = \gamma_i(1); \quad 1 \leq i \leq N$$
(2.26)

New estimates of the model parameters are calculated with the expressions:

$$\hat{a}_{ij} = \frac{\sum_{t=1}^{T} \xi_t(i,j)}{\sum_{t=1}^{T} \gamma_i(t)}; \quad 1 \leq i, j \leq N \tag{2.27}$$

$$\hat{b}_j(k) = \frac{\sum_{t=1}^{T} |_{O_t=k} \gamma_t(j)}{\sum_{t=1}^{T} \gamma_j(t)}; \quad 1 \leq j \leq N \text{ and } 1 \leq k \leq K \tag{2.28}$$

2.7 Exercises

(1) The following is a sequence of observations on whether the weather is sunny or rainy.

   (a) What is the probability that it is sunny today if it was sunny the day before?
   (b) What is the probability that it is rainy today if it was sunny the day before?
   (c) What is the probability that it is rainy today if it was rainy the day before?
   (d) What is the probability that it is sunny today if it was rainy the day before?

(2) Draw and label the hidden Markov model given in Question 1
(3) Write the transition probability matrix
(4) Assume that Ireh's mood is dependent on the weather and over a period of eleven days her mood is as shown. Compute the emission probabilities based on her mood.

(5) Draw the HMM showing the transition and emission probabilities. Label the diagram with the probabilities
(6) Write the emission probability matrix
(7) Consider a random day. What is the probability that the day is sunny or rainy?

3

Introduction to Kalman Filters

3.1 Introduction

Kalman filters (KF) remain a unique algorithm which has found use repeatedly in tracking problems in system dynamics and yet remains one of the least understood topic in data handling techniques. Its power is in its iterative calculations which allow a user to input a new reading and use it to provide an estimate of parameters of underlying process. In object tracking, for example, the underlying variables to be estimated include the velocity, position and acceleration of a moving object like an aircraft, bullet or space vehicle. It is most useful when there are uncertainties in the measured values and the true value is desired.

An example of interest is in the measurement of temperature using a temperature sensor. The sensor provides periodically temperature reading, which has uncertainties or errors in the measurement. The objective is to find the correct or true temperature.

3.2 Scalar Form

Two types of Kalman filters, the scalar form and the matrix form are described in this chapter. The Kalman filter is first introduced using a simple scalar example in which only one variable is to be estimated. The example is that of temperature measurement by a sensor. The actual signal being measured does not really matter as it could be anything like humidity, pressure or wind speed.

Many processes result to noisy data. The true data is buried in noise. The desire is to estimate the actual data using an iterative approach. Each new measurement is used to estimate what the true value of the data should have been if there were no noise. Therefore, it can prove very useful in tracking the motion of objects and the dynamic behaviour of processes. The measured value contains some errors, or uncertainties which can be used to quickly

35

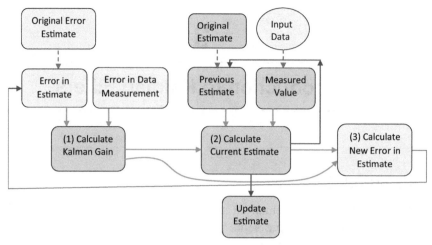

Figure 3.1   Kalman filter processing steps.

compute what the real value should be. For that reason, we look at it as a method of 'filtering' out uncertainties or errors in measurements, hence the name "Kalman Filter". Although it is not a real filter in the sense of filters in digital signal processing, but it possesses the characteristics of filters – that of giving a true value of a signal or data.

Figure 3.1 shows three processing steps in Kalman filtering. In step (1), the Kalman filter gain is computed. It has as input error in the estimate and error (uncertainty) in the data measurement. The Kalman gain provides a measure of the relative importance of the previous error and the uncertainty in the measurement. During the first iteration, we use 'original error' estimate instead of the so-called 'previous error'. Once that is done, we do not go back to that original error ever again. It is vital to realise that this 'original error' could be any reasonable value chosen to start the process.

The calculated Kalman gain is used as input to step (2) which calculates the current estimate of the data. This step also has two other inputs, the previous estimate and measured value. In this step, the current estimate is computed. For the first iteration, we use an 'original estimate'. The choice of this value is rather arbitrary and does not matter as the system quickly converges to the current estimate after a few iterations. Once we calculate the current estimate, it becomes the previous estimate for the next iteration. Notice from the diagram that the data keeps coming in for all the iterations to follow.

In step (3), we calculate the new error in the estimate using the current Kalman filter gain and the output of step (2), which is the current estimate.

Once the new error is calculated, it is feedback to the input and it becomes the error in the estimate for the next iteration.

Each time we go through the iteration an output result is issued and becomes what we 'update' the 'estimate' (update estimate in the diagram). It is used to update the value of the process for this step of the iteration. This really is the objective of the Kalman filtering to iteratively get new update estimates. For example, if the process in question is measuring temperature, the 'update estimate' is a value of temperature, which we think is getting closer and closer to the true value of the temperature. If our process is a radar tracking system tracking an aircraft, the updated estimate gets closer and closer to the true position of the aircraft. This may also include its velocity and altitude or acceleration.

3.2.1 Step (1): Calculate Kalman Gain

Consider the following three block diagrams in Figure 3.2, which are essential for estimating the Kalman gain.

Kalman Gain: Figure 3.2 is an illustration of factors that influence the value of the Kalman gain. It shows two inputs. The Kalman gain is calculated using the error in the estimate ε_{est} and error or uncertainty in the measurement ε_{mes}. It is calculated as a ratio of these errors using Equation (3.1).

$$K = \frac{\varepsilon_{est}}{\varepsilon_{est} + \varepsilon_{mes}} \tag{3.1}$$

From Equation (3.1) it is obvious that the Kalman gain satisfies the inequality $0 \leq K \leq 1$. It is a positive number less than one. When the Kalman gain tends to one $(K \rightarrow 1)$, the measurements become more and more accurate. This also means that for large Kalman filter gain, the error in the estimate dominates the gain and the error in measurement is small. The estimates on the other hand become unstable.

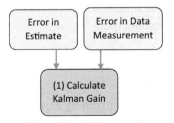

Figure 3.2 Kalman gain.

When also the Kalman gain tends to zero $K \to 0$, the measurements become more and more inaccurate. The estimates in this case become more and more stable and have smaller and smaller errors in them. The error in measurement is then very large. This can be seen from Equation (3.1).

Estimate of State: Let E_k be the current estimate, E_{k-1} be the previous estimate and χ be the measurement. Hence, the current estimate is obtained using the following equation:

$$E_k = E_{k-1} + K(\chi - E_{k-1}) \tag{3.2}$$

From both Equations (3.1) and (3.2), we note that large Kalman filter gain means that the update to the previous estimate is large.

Error in Estimate: Let the error in the estimate at time k be denoted with the variable $\varepsilon_{est}(k)$ and the error in the estimate at time $k-1$ be $\varepsilon_{est}(k-1)$. Hence, the error in the estimate at time t can be written in terms of the Kalman gain with the expression:

$$\varepsilon_{est}(k) = \left(\frac{\varepsilon_{mes}\varepsilon_{est}(k-1)}{\varepsilon_{est}(k-1) + \varepsilon_{mes}} \right) \tag{3.3}$$

Equation (3.3) can be simplified to

$$\varepsilon_{est}(k) = (1 - K)\varepsilon_{est}(k-1) \tag{3.4}$$

From Equation (3.4), if the Kalman gain is large, the error in the measurement is small and the current error in the estimate is also small. However, if the Kalman gain is very small, it means the error in the measurement is very large. This will mean also that the iteration takes longer to converge.

Example: A sensor was deployed on the river Nile to measure the depth of water with the objective of determining when flood could occur. The true depth of the river is 72 m. The initial estimate of depth is 68 m. The error in the estimate is 2 m. If the initial measurement of depth is 75 m and the sensor has an error of 4 m, estimate the

(i) Kalman gain
(ii) Current estimated error
(iii) Error in the estimate

Solution: The Kalman gain is given by Equation (3.1).

(i) The Kalman gain is

$$\varepsilon_{est} = 2; \quad \varepsilon_{mes} = 4$$

$$K = \frac{\varepsilon_{est}}{\varepsilon_{est} + \varepsilon_{mes}} = \frac{2}{2+4} = \frac{1}{3}$$

(ii) The current water level estimate is obtained from Equation (3.2), which is

$$E_k = E_{k-1} + K(\chi - E_{k-1}) = 68 + \frac{1}{3}(75 - 68) = 70.33 \text{ m}$$

(ii) The error in estimate is given by Equations (3.3) or (3.4). We will use Equation (3.4). Note the initial estimate error was 2 m; therefore, the new error in estimate is

$$\varepsilon_{est}(k) = (1 - K)\varepsilon_{est}(k - 1) = \left(1 - \frac{1}{3}\right) \times 2 = \frac{4}{3}$$

Exercise: A water level sensor was used to measure water levels in River Nile resulting in the following values in the following table. In Example 1, we have computed the first set of results. Complete the rest of the calculations for Kalman gain, estimates of water levels and their estimated errors. The values in red are given in the problem statement. Notice in the given problem, the sensor error remains constant as long as we continue to use the same sensor.

	χ (Measurement)(m)	ε_{mes} (measurement Error)	E_k(m) Estimate	$\varepsilon_{est}(k-1)$ (Previous Estimate Error)	K Kalman Gain	$\varepsilon_{est}(k)$ Current Estimate Error
k−1			68	2		
k	75	4	70.33		1/3	4/3
k+1	74	4				
k+2	71.5	4				
k+3	73	4				

3.3 Matrix Form

In this section, we consider a more rigorous example of Kalman filtering in which matrices play a part. Most applications of Kalman filtering will require this form in which the input is a vector, the system model is a matrix and the estimated errors are covariance matrices. The analysis in this chapter follows the form given by Michel van Biezen [1].

The basis of Kalman filters stem from the following. Assume there is a process whose dynamic equations are known. There is noise in the process. The manifestation of this process is measured using a sensor. There is also measurement noise. The role of the Kalman filter is to determine the true nature of the output of the process using the noisy measurements. Consider a typical example, a drone being tracked with a radar system. The radar provides observations of the position, velocity and acceleration of the drone. These variables determine the measured position of the drone. The drone model (process) also provides an estimate of the drone position. Our objective is to minimize the error between the output of the process and measurement (Figure 3.3) and provide a good knowledge of the system.

Consider a process called the model in Figure 3.4. This process (model) has dynamic equation

$$x_t = A_t x_{t-1} + B_t u_t + w_t \qquad (3.5)$$

- x_t is the state vector at time t. It contains system variables (e.g., coordinates, velocity, acceleration, etc.)

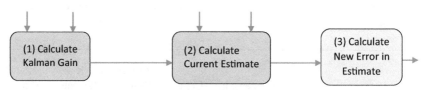

Figure 3.3   Three stages of Kalman filter computation.

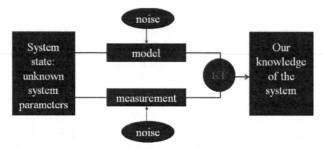

Figure 3.4   Kalman filter in a nutshell.

- A_t is the $n \times n$ state transition matrix containing descriptions of how its velocity, acceleration and coordinates at time $t-1$ affect its position at time t
- B_t is a matrix describing how control input changes (e.g., in acceleration, velocity, etc.) affect the system state
- u_t is a vector containing the control variables (turning, throttle and breaking)
- w_t is the process noise vector
- Q is the covariance matrix of the process. The process noise is multivariate and has normal distribution with zero mean

The measurement equation provides how the sensor acquires the state of the process.

$$y_t = H_t x_t + v_t \tag{3.6}$$

- y_t is the measurement vector
- H_t is the measurement transformation matrix, transforming the state into the measurement vector
- v_t is the measurement noise vector
- R is the measurement noise covariance matrix. The measurement noise is drawn from a Gaussian white noise process with zero mean

Equations (3.5) and (3.6) provide the full descriptions of the underlying process and measurements undertaken using sensors. Figure 3.5 illustrates how these processes and measurement models are combined to provide a Kalman filter output that is more representative of the true nature of the system.

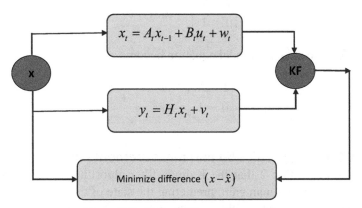

Figure 3.5   Kalman filter system model.

3.3.1 Models of the State Variables

In this section, we provide examples of the state, process, measurement and noise variables described in the previous section using examples.

To illustrate the formulation provided in this section, tracking of a drone is undertaken. We assume the drone is travelling on a straight line at constant speed. There is neither vertical speed nor acceleration. During the driving process, the drone pilot applies a force in the forward direction or breaking to slow it down. In the forward direction, the force is proportional to the mass of the drone m resulting to the control expression

$$\mathbf{u}_t = \frac{f_t}{m} \qquad (3.7)$$

We describe the linear motion of the drone with the expression

$$x_t = x_{t-1} + \dot{x}_{t-1} \times \Delta t + \frac{(\Delta t)^2 f_t}{2\,m} \qquad (3.8)$$

$$\dot{x}_t = \dot{x}_{t-1} + \frac{\Delta t \cdot f_t}{m}$$

The state vector for the drone is therefore defined by two variables, the position and the velocity of the drone. This is

$$\mathbf{x}_t = [x_t, \dot{x}_t]^T \qquad (3.9)$$

The transition matrix is given by the expression

$$A = \begin{bmatrix} 1 & \Delta t \\ 0 & 1 \end{bmatrix} \quad \text{and} \quad B = \begin{bmatrix} \frac{(\Delta t)^2}{2} \\ \Delta t \end{bmatrix}$$

Therefore, the system model equation becomes:

$$\begin{bmatrix} x_t \\ \dot{x}_t \end{bmatrix} = \begin{bmatrix} 1 & \Delta t \\ 0 & 1 \end{bmatrix} \cdot \begin{bmatrix} x_{t-1} \\ \dot{x}_{t-1} \end{bmatrix} + \begin{bmatrix} \frac{(\Delta t)^2}{2} \\ \Delta t \end{bmatrix} \cdot \frac{f_t}{m} \qquad (3.10)$$

3.3.1.1 Using prediction and measurements in Kalman filters

We saw in Section 2 that the Kalman filter algorithm involves three steps, calculating the Kalman gain, predicting the state of the process and updating the measurement. These three steps are derived for the matrix case. When there is measurements and prediction of the value as well, the objective is to minimize the error between measurement and prediction as much as possible. This requires minimizing the error through covariance analysis. The process equation is often written in a form which shows that the prediction is based on prior events or predictions. This is shown in the equation

$$\hat{\mathbf{x}}_{t|t-1} = A_t \hat{\mathbf{x}}_{t-1|t-1} + B_t u_t \tag{3.11}$$

The error covariance matrix P is also given by the expression

$$P_{t|t-1} = A_t P_{t-1|t-1} A_t^T + Q_t \tag{3.12}$$

where Q is the process noise. To derive Equation (3.1), we first subtract Equation (3.10) from Equation (3.5), which means we subtract the predicted value from the actual value to form the error. Then, we evaluate the expectation of the error. We will henceforth drop the second set of subscripts in Equations (3.10) and (3.11) and retain only one subscript on each variable. This is purely to reduce confusion in subscripts and understanding. Therefore, the expectation of the error is

$$P_{t|t-1} = E\left[(\mathbf{x}_t - \hat{\mathbf{x}}_{t|t-1})(\mathbf{x}_t - \hat{\mathbf{x}}_{t|t-1})^T\right] \tag{3.13}$$

$$e_{t|t-1} = A(\mathbf{x}_{t-1} - \hat{\mathbf{x}}_{t|t-1}) + w_t \tag{3.14}$$

The expectation of this error

$$P_{t|t-1} = E[A(\mathbf{x}_{t-1} - \hat{\mathbf{x}}_{t|t-1}) + w_t] \cdot [A(\mathbf{x}_{t-1} - \hat{\mathbf{x}}_{t|t-1}) + w_t]^T$$

$$= AE[(\mathbf{x}_{t-1} - \hat{\mathbf{x}}_{t|t-1})^T \cdot (\mathbf{x}_{t-1} - \hat{\mathbf{x}}_{t|t-1}) A^T + A(\mathbf{x}_{t-1} - \hat{\mathbf{x}}_{t|t-1})^T \cdot w_t$$

$$+ w_t \cdot (\mathbf{x}_{t-1} - \hat{\mathbf{x}}_{t|t-1}) A^T] + E[w_t^T \cdot w_t] \tag{3.15}$$

The process noise and state estimation errors are uncorrelated. Therefore, the terms involving them can be set to zero.

$$E[(\mathbf{x}_{t-1} - \hat{\mathbf{x}}_{t|t-1})^T \cdot w_t] = E[w_t^T \cdot (\mathbf{x}_{t-1} - \hat{\mathbf{x}}_{t|t-1})] = 0 \tag{3.16}$$

Therefore, the expectation of the error simplifies to the expression

$$P_{t|t-1} = AE[(\mathbf{x}_{t-1} - \hat{\mathbf{x}}_{t|t-1})^T \cdot (\mathbf{x}_{t-1} - \hat{\mathbf{x}}_{t|t-1})] \cdot A^T + E[w_t^T \cdot w_t] \tag{3.17}$$

Since

$$P_{t-1|t-1} = E[(\mathbf{x}_{t-1} - \hat{\mathbf{x}}_{t|t-1})^T \cdot (\mathbf{x}_{t-1} - \hat{\mathbf{x}}_{t|t-1})] \text{ and } E[w_t^T \cdot w_t] = Q_t$$

$$P_{t|t-1} = A P_{t|-1t-1} A^T + Q_t \tag{3.18}$$

The measurement update step involves the use of the error probability, which we derived in the previous step. To update the measurement, the following expression is used in addressing the fact that our prediction of the value of the process is not accurate. It has errors. This error can be used again combined with the Kalman gain to improve on our prediction in the previous time step. This is done using the measurement update equation:

$$\hat{\mathbf{x}}_{t|t} = \hat{\mathbf{x}}_{t|t-1} + \mathbf{K_t}(\mathbf{z_t} - \mathbf{H_t}\hat{\mathbf{x}}_{t|t-1}) \tag{3.19}$$

$$\mathbf{P_{t|t}} = \mathbf{P_{t|t-1}} - \mathbf{K_t H_t P_{t|t-1}} = \mathbf{P_{t|t-1}}(\mathbf{I} - \mathbf{K_t H_t}) \tag{3.20}$$

From Equation (3.19), the Kalman filter gain can be obtained. These expressions are derived in the next section. The term in bracket in Equation (3.19) is the error between the measurement and the predicted value of the process at time t. This error is used to update the value of the estimate at time $t - 1$ to give the value of estimate at time t. As this error gets smaller and smaller, the value of the process becomes more and more realistic or accurate.

In terms of tracking the drone in our example, the estimate of its position and velocity based on the device on the drone and measurement by a beacon being used to track the system are combined to provide the best information on where the drone actually is.

3.3.2 Gaussian Representation of State

The initial position of the drone is assumed to have a Gaussian probability density function at time $t = 0$. When the drone moves to a new location at time $t = 1$, it is also assumed that the position and velocity of the drone is still Gaussian in distribution [2].

At the first location, the best estimate of the position is given by $\hat{x}_1 = z_1$ (Figure 3.6, green colour) with standard deviation (uncertainty) σ_1 and variance σ_1^2. The distribution at this position is thus described by the

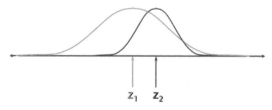

Figure 3.6   Gaussian distribution [2].

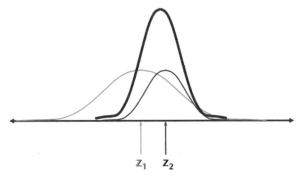

Figure 3.7 Fused Gaussian distributions.

Gaussian function

$$z_1(r; \mu_1, \sigma_1) = \frac{1}{\sqrt{2\pi\sigma_1^2}} e^{\frac{-(r-\mu_1)^2}{2\sigma_1^2}} \tag{3.21}$$

The second measurement $\hat{x}_2 = z_2$ (brown) has variance σ_2^2 and uncertainty σ_2 and with similar distribution as the first.

$$z_2(r; \mu_2, \sigma_2) = \frac{1}{\sqrt{2\pi\sigma_2^2}} e^{\frac{-(r-\mu_2)^2}{2\sigma_2^2}} \tag{3.22}$$

By fusing the information provided by the two probability distribution functions, the pink distribution is obtained. The sum of two Gaussian functions is also Gaussian and hence the fused system has distribution [2]

$$z(r; \mu_1, \sigma_1, \mu_2, \sigma_2) = \frac{1}{\sqrt{2\pi\sigma_1^2}} e^{\frac{-(r-\mu_1)^2}{2\sigma_1^2}} + \frac{1}{\sqrt{2\pi\sigma_2^2}} e^{\frac{-(r-\mu_2)^2}{2\sigma_2^2}}$$

$$= \frac{1}{2\pi\sqrt{\sigma_1^2\sigma_2^2}} e^{-\left(\frac{(r-\mu_1)^2}{2\sigma_1^2} + \frac{(r-\mu_2)^2}{2\sigma_2^2}\right)} \tag{3.23}$$

The best estimate of the position is a weighted average given by

$$\hat{x}_2 = \frac{\frac{1}{\sigma_1^2}z_1 + \frac{1}{\sigma_2^2}z_2}{\frac{1}{\sigma_1^2} + \frac{1}{\sigma_2^2}} = \hat{x}_1 + \frac{\sigma_1^2}{\sigma_1^2+\sigma_2^2}(z_2 - \hat{x}_1) \tag{3.24}$$

Therefore, we can rewrite the exponent of the fused pdf as:

$$\mu_z = \frac{\sigma_2^2\mu_1 + \sigma_1^2\mu_2}{\sigma_1^2 + \sigma_2^2} = \mu_1 + \frac{\sigma_1^2}{\sigma_1^2+\sigma_2^2}(\mu_2 - \mu_1) \tag{3.25}$$

Also

$$\sigma_z^2 = \frac{\sigma_1^2 \sigma_2^2}{\sigma_1^2 + \sigma_2^2} = \sigma_1^2 - \frac{\sigma_1^4}{\sigma_1^2 + \sigma_2^2} \tag{3.26}$$

These two equations represent the measurement update. The uncertainty in the best estimate is therefore

$$\hat{\sigma}_2^2 = \frac{\sigma_1^2 \sigma_2^2}{\sigma_1^2 + \sigma_2^2} \tag{3.27}$$

Therefore, the probability density function for the fused position is

$$z(r; \mu_z, \sigma_z) = \frac{1}{\sqrt{2\pi\sigma_z^2}} e^{-\left(\frac{(r - \mu_z)^2}{2\sigma_z^2}\right)} \tag{3.28}$$

These equations are suitable for situations in which the prediction and measurements are undertaken in the same domain. If, for example, the distance and velocity of a moving target are measured in different units (metres and seconds for example), it is essential to use a transformation which maps them into the same domain. Hence, the probability density functions for the prediction and measurements can be multiplied together since they are in the same domain. For example, in a drone tracking system, the prediction can be converted to the same as the measurement domain by converting the time of flight to distance by dividing distance with velocity. This section illustrates that form of approach in Kalman filtering.

Therefore, by letting t represent the time axis, we can rewrite the previous equations as:

$$z_1(t; \mu_1, \sigma_1, c) = \frac{1}{\sqrt{2\pi \left(\frac{\sigma_1}{c}\right)^2}} e^{-\frac{\left(t - \frac{\mu_1}{c}\right)^2}{2\left(\frac{\sigma_1}{c}\right)^2}} \tag{3.29}$$

$$z_2(t; \mu_2, \sigma_2) = \frac{1}{\sqrt{2\pi\sigma_2^2}} e^{\frac{-(t - \mu_2)^2}{2\sigma_2^2}} \tag{3.30}$$

Let $H = \frac{1}{c}$ and [2] Kalman Gain $K = \frac{H\sigma_1^2}{H^2\sigma_1^2 + \sigma_2^2}$
Then

$$\mu_z = \mu_1 + K(\mu_2 - H\mu_1) \tag{3.31}$$

In a similar manner, the fused variance becomes

$$\sigma_{fused}^2 = \sigma_1^2 - \left(\frac{\left(\frac{\sigma_1}{c}\right)^2}{\left(\frac{\sigma_1}{c}\right)^2 + \sigma_2^2}\right) \cdot \left(\frac{\sigma_1}{c}\right)^2 = \sigma_1^2 - KH\sigma_1^2 \tag{3.32}$$

With this model, the following comparison with the linear model can be drawn.

The Kalman gain is thus equivalently written as

$$\mathbf{K}_t = \mathbf{P}_{t|t-1}\mathbf{H}_t^T(\mathbf{H}_t\mathbf{P}_{t|t-1}\mathbf{H}_t^T + \mathbf{R}_t)^{-1} \tag{3.33}$$

We have now shown how the Kalman gain parameters map in different coordinate systems to the same coordinate system. Next we show how the covariance matrix also maps in the same coordinate systems.

The fused variables relate to both the state updating procedure and how the covariance matrix is updated. The state update expression is first presented based on the mapping in Table 3.1. We have shown that

$$\mu_z = \mu_1 + K(\mu_2 - H\mu_1) = \mu_1 + \frac{H\sigma_1^2}{H^2\sigma_1^2 + \sigma_2^2}(\mu_2 - H\mu_1) \tag{3.34}$$

From Table 3.1, this equation can be written as

$$\hat{\mathbf{x}}_{t|t} = \hat{\mathbf{x}}_{t|t-1} + \mathbf{K}_t(\mathbf{z}_t - \mathbf{H}_t\hat{\mathbf{x}}_{t|t-1}) \tag{3.35}$$

Similarly, to update the covariance matrix, the expressions involved from the same coordinate system to different coordinate systems is

$$\sigma_z^2 = \sigma_1^2 - \frac{H\sigma_1^2}{H^2\sigma_1^2 + \sigma_2^2}H\sigma_1^2 \tag{3.36}$$

Using the mapping in Table 3.1, we have

$$\mathbf{P}_{t|t} = \mathbf{P}_{t|t-1} - \mathbf{K}_t\mathbf{H}_t\mathbf{P}_{t|t-1} \tag{3.37}$$

Table 3.1 Comparison of Kalman filter parameters in same and different coordinate systems [2]

Description of Variable	Comparison
The state vector (prediction) before data fusion	$\hat{\mathbf{x}}_{t|t-1} \Rightarrow \mu_1$
After data fusion, the state vector	$\hat{\mathbf{x}}_{t|t} \Rightarrow \mu_z$
The measurement vector	$\mathbf{z}_t \Rightarrow \mu_2$
The state covariance matrix before data fusion	$\mathbf{P}_{t|t-1} \Rightarrow \sigma_1^2$
After data fusion, the state covariance matrix	$\mathbf{P}_{t|t} \Rightarrow \sigma_z^2$
The measurement error matrix	$\mathbf{R}_t \Rightarrow \sigma_2^2$
The transformation matrix which maps the state vector variables to the measurement domain	$\mathbf{H}_t \Rightarrow H$
The Kalman Gain	$\mathbf{K}_t \Rightarrow K = \frac{H\sigma_1^2}{H^2\sigma_1^2+\sigma_2^2}$

Thus, the distinction between Kalman filter in the same coordinate system and in different coordinate system is the requirement for a mapping of the Kalman gain, state update and covariance matrix to the different coordinate system. The $\mathbf{H_t}$ matrix is used for the transformation mapping.

3.4 The State Matrix

We have laid significant foundations for understanding Kalman filters. In this section, we add to the foundations by providing examples of how to compose the state matrix for exemplary systems. We will give several examples. The general state equation is the same as given before and is

$$X_k = AX_{k-1} + Bu_k + w_k \qquad (3.38)$$

Notice the use of k to represent the current time as is common in Kalman filter operations. The rest of the variables retain their meanings as in Equation (3.5).

3.4.1 State Matrix for Object Moving in a Single Direction

For objects moving in the x-direction only, we are interested in the position and velocity of the object. We assume the object has zero acceleration. Therefore, the state matrix has two variables:

$$X = \begin{bmatrix} x \\ \dot{x} \end{bmatrix} = \begin{bmatrix} x \\ v \end{bmatrix}$$

$$v = \frac{dx}{dt} \qquad (3.39)$$

For this case, the state matrix is a vector (a degenerate matrix with only one column). The equation of motion is thus

$$x = x_0 + \dot{x} \cdot t \qquad (3.40)$$

Another single-dimension case is when an object moves in the y-direction like a falling object. This example is that of a falling object such as a stone or fruit falling from a tree. We assume there are no cross-winds and it falls freely. The object does not experience any acceleration. The state matrix is also given by the equations

$$X = \begin{bmatrix} y \\ \dot{y} \end{bmatrix} = \begin{bmatrix} y \\ v \end{bmatrix}$$

$$v = \frac{dy}{dt} \tag{3.41}$$

The equation of motion in the y-direction is thus $y = y_0 + \dot{y} \cdot t$. The state is updated at discrete times Δt, which leads to the A matrix of the form

$$A = \begin{bmatrix} 1 & \Delta t \\ 0 & 1 \end{bmatrix} \tag{3.42}$$

and when the object is moving in the x-direction and identically the relevant equation is

$$AX = \begin{bmatrix} 1 & \Delta t \\ 0 & 1 \end{bmatrix} \begin{bmatrix} x \\ \dot{x} \end{bmatrix} \tag{3.43a}$$

When the object is moving in the y-direction (e.g., a falling stone or fruit, or water rising in a tank).

$$AX = \begin{bmatrix} 1 & \Delta t \\ 0 & 1 \end{bmatrix} \begin{bmatrix} y \\ \dot{y} \end{bmatrix} \tag{3.43b}$$

Let us consider the case of a falling object further. A falling object will experience acceleration due to gravity. Hence, we have the B matrix coming into play in the state equation as well. This gives the B matrix as

$$B = \begin{bmatrix} \frac{(\Delta t)^2}{2} \\ \Delta t \end{bmatrix} \tag{3.44}$$

The acceleration experienced by the falling object is due to gravity and $u = [g]$. Hence, the state equation so far can be written as

$$X_k = AX_{k-1} + Bu_k = \begin{bmatrix} 1 & \Delta t \\ 0 & 1 \end{bmatrix} \begin{bmatrix} y \\ \dot{y} \end{bmatrix} + \begin{bmatrix} \frac{(\Delta t)^2}{2} \\ \Delta t \end{bmatrix} [g] \tag{3.45}$$

This means

$$X_k = \begin{bmatrix} y + \Delta t \cdot \dot{y} \\ \dot{y} \end{bmatrix} + \begin{bmatrix} g \frac{(\Delta t)^2}{2} \\ g\Delta t \end{bmatrix} = \begin{bmatrix} y + \Delta t \cdot \dot{y} + g \frac{(\Delta t)^2}{2} \\ \dot{y} + g\Delta t \end{bmatrix} \quad (3.46)$$

Remember that the next time step we must reuse the expression $X_k = AX_{k-1} + Bu_k$, where the above X_k now becomes X_{k-1} for the next iteration or update. Notice that if there were no acceleration like in the case of rising water in a tank $a = 0$ and Bu $= 0$ for that case.

Exercise: *What is the equivalent state equation for an object moving in the x-direction with acceleration* $u = [a]$?

For the general iterative case of the above equation, the state equation is re-written in the form

$$X_k = AX_{k-1} + Bu_k = \begin{bmatrix} 1 & \Delta t \\ 0 & 1 \end{bmatrix} \begin{bmatrix} y_{k-1} \\ \dot{y}_{k-1} \end{bmatrix} + \begin{bmatrix} \frac{(\Delta t)^2}{2} \\ \Delta t \end{bmatrix} [g]$$

When the matrices are multiplied together, we have the same form of solution but this time with the correct k subscripts as

$$X_k = \begin{bmatrix} y_{k-1} + \Delta t \cdot \dot{y}_{k-1} \\ \dot{y}_{k-1} \end{bmatrix} + \begin{bmatrix} g \frac{(\Delta t)^2}{2} \\ g\Delta t \end{bmatrix} = \begin{bmatrix} y_{k-1} + \Delta t \cdot \dot{y}_{k-1} + g \frac{(\Delta t)^2}{2} \\ \dot{y}_{k-1} + g\Delta t \end{bmatrix}$$

The current position of the object is the sum of the initial position plus two terms, the first correction due to velocity and the second due to acceleration of the object. The velocity of the object is equally updated with the correction due to acceleration. The dynamics of the falling object is thus $y = y_0 + \dot{x} \cdot t + \dot{y} \cdot t$.

Let the initial position of the object be $y_{k-1} = 50$ m, $\Delta t = 1$ sec, initial velocity $\dot{y}_{k-1} = 0$ m/sec and $g = -9.8$ m/sec^2. Therefore, the new position of the falling object can now be computed to be

$$X_k = \begin{bmatrix} 50 + 0 - \frac{9.8}{2} \\ 0 - 9.8 \end{bmatrix} = \begin{bmatrix} 45.1 \\ -9.8 \end{bmatrix}$$

Therefore, the new state shows that the new position of the object is 45.1 m and the velocity is -9.8 m/sec^2.

Example: *A moving military tank is being tracked by a radar system on ground. If the speed of the tank is 10 m/sec, acceleration of the tank is* constant at 5 m/sec² square and initial position is 60 m, track this vehicle for three consecutive seconds.

Solution:

$$X_k = \begin{bmatrix} x_{k-1} + \Delta t \cdot \dot{x}_{k-1} + a_x \frac{(\Delta t)^2}{2} \\ \dot{x}_{k-1} + a_x \Delta t \end{bmatrix}$$

At $k = 0$, the state of the tank is $X_0 = \begin{bmatrix} 60 \\ 10 \end{bmatrix}$

$$k = 1, X_1 = \begin{bmatrix} x_0 + \Delta t \cdot \dot{x}_0 + a_x \frac{(\Delta t)^2}{2} \\ \dot{x}_0 + a_x \Delta t \end{bmatrix} = \begin{bmatrix} 60 + (1)(10) + 5\frac{1^2}{2} \\ 10 + (5)(1) \end{bmatrix}$$

$$= \begin{bmatrix} 72.5 \\ 15 \end{bmatrix} = \begin{bmatrix} x_1 \\ \dot{x}_1 \end{bmatrix};$$

$$k = 2, X_2 = \begin{bmatrix} x_1 + \Delta t \cdot \dot{x}_1 + a_x \frac{(\Delta t)^2}{2} \\ \dot{x}_1 + a_x \Delta t \end{bmatrix} = \begin{bmatrix} 72.5 + (1)(15) + 5\frac{1^2}{2} \\ 15 + (5)(1) \end{bmatrix}$$

$$= \begin{bmatrix} 90 \\ 20 \end{bmatrix} = \begin{bmatrix} x_2 \\ \dot{x}_2 \end{bmatrix}$$

$$k = 3, X_3 = \begin{bmatrix} x_2 + \Delta t \cdot \dot{x}_2 + a_x \frac{(\Delta t)^2}{2} \\ \dot{x}_2 + a_x \Delta t \end{bmatrix} = \begin{bmatrix} 90 + (1)(20) + 5\frac{1^2}{2} \\ 20 + (5)(1) \end{bmatrix}$$

$$= \begin{bmatrix} 112.5 \\ 25 \end{bmatrix} = \begin{bmatrix} x_3 \\ \dot{x}_3 \end{bmatrix}$$

Exercise: Given the dynamics of the moving tank as $x = x_0 + \dot{x} \cdot t + \frac{\ddot{x}}{2} \cdot t^2$

(i) Prove that the position of the tank after 3 seconds is equal to the position obtained with the Kalman filter method.

(ii) What is the velocity of the tank after 3 seconds?

Solution: Given, $t = 3, x_0 = 60$ m; $\dot{x} = 10$ m/s; $\ddot{x} = 5$ m/s²

(i) The position of the tank after 3 seconds

$$x = x_0 + \dot{x} \cdot t + \frac{\ddot{x}}{2} \cdot t^2 = 60 + (10)(3) + \frac{5}{2}(3)^2$$
$$= 60 + 30 + 22.5 = 112.5 \text{ m}$$

(ii) The velocity of the tank is $\dot{x} = \dot{x}_0 + \ddot{x} \cdot t = 10 + (5)(3) = 25$ m/s

3.4.1.1 Tracking including measurements

Suppose in addition to the state equation we also have a measurement model which describes how a sensor measures the position and velocity of the falling object. This inclusion of measurement changes the system equations to be

$$\left. X_k = \begin{matrix} AX_{k-1} + Bu_k + w_k \\ Y_k = C \cdot X_k + z_k \end{matrix} \right\} \tag{3.47}$$

The second equation is called the measurement equation and z_k is measurement noise. For the example of falling object, assume the sensor is also able to measure position and velocity; therefore, a suitable C matrix would be a 2×2 matrix with zero measurement noise to give the expression

$$Y_k = C \cdot X_k + z_k = \begin{bmatrix} 1 & 0 \\ 0 & 1 \end{bmatrix} \begin{bmatrix} y'_k \\ \dot{y}'_k \end{bmatrix} + 0 \tag{3.48}$$

If the sensor had measured position only, then $C = [1 \ 0]$. If however the sensor measured velocity only and not position, then $C = [0 \ 1]$.

3.4.2 State Matrix of an Object Moving in Two Dimensions

Consider a drone moving on a plane (a two-dimensional situation). Its state matrix therefore will contain variations in x, y and velocities in these two coordinate axes. Thus, the state matrix is

$$X = \begin{bmatrix} x \\ \dot{x} \\ y \\ \dot{y} \end{bmatrix} \quad \text{or} \quad X = \begin{bmatrix} x \\ y \\ \dot{x} \\ \dot{y} \end{bmatrix} \tag{3.49}$$

The order in which the variables in the state matrix are written does not matter provided, we remain consistent once we start with a particular order. This drone has both positions and velocities in the two-coordinate system. The equation of motion is also a combination of these variables as

$$h = x_0 + y_0 + \dot{x} \cdot t + \dot{y} \cdot t \tag{3.50}$$

The matrices which come into play for the two-dimensional case are a bit different from the single-dimension matrices. The matrices are also based on the way the state vector variables are arranged. In this book, we choose the form where we first show the coordinates of the position of the object and then the velocities in the two dimensions. Thus, for the state vector

$$X = \begin{bmatrix} x \\ y \\ \dot{x} \\ \dot{y} \end{bmatrix} \quad \text{the matrix } A = \begin{bmatrix} 1 & 0 & \Delta t & 0 \\ 0 & 1 & 0 & \Delta t \\ 0 & 0 & 1 & 0 \\ 0 & 0 & 0 & 1 \end{bmatrix} \text{ and}$$

$$AX = \begin{bmatrix} 1 & 0 & \Delta t & 0 \\ 0 & 1 & 0 & \Delta t \\ 0 & 0 & 1 & 0 \\ 0 & 0 & 0 & 1 \end{bmatrix} \cdot \begin{bmatrix} x \\ y \\ \dot{x} \\ \dot{y} \end{bmatrix}$$

(3.51)

Exercise 2: Write the expression for A and AX when the state vector $A = \begin{bmatrix} x & \dot{x} & y & \dot{y} \end{bmatrix}^T$

Solution:

$$AX = \begin{bmatrix} 1 & \Delta t & 0 & 0 \\ 0 & 1 & 0 & 0 \\ 0 & 0 & 1 & \Delta t \\ 0 & 0 & 0 & 1 \end{bmatrix} \cdot \begin{bmatrix} x \\ \dot{x} \\ y \\ \dot{y} \end{bmatrix}$$

It is now essential to look at the control part of the state equation. For the two-dimensional case for an object falling from a height, it has acceleration in the x- and y-directions with which we can update both the position of the object and its velocity. It is essential to match the format of the state vector (4×1 in our case) since we are updating the state vector eventually. The control equation therefore need to be written and is

$$Bu_k = \begin{bmatrix} \frac{(\Delta t)^2}{2} & 0 \\ 0 & \frac{(\Delta t)^2}{2} \\ \Delta t & 0 \\ 0 & \Delta t \end{bmatrix} \cdot \begin{bmatrix} a_x \\ a_y \end{bmatrix} = \begin{bmatrix} \frac{(\Delta t)^2 a_x}{2} \\ \frac{(\Delta t)^2 a_y}{2} \\ \Delta t \cdot a_x \\ \Delta t \cdot a_y \end{bmatrix}$$

The result is thus a 4×1 matrix or vector. The first two elements have unit of distance and the last two elements have unit of velocity. Therefore, they are suitable for use in updating the state vector as written initially. Therefore, the state is

$$X_k = AX_{k-1} + Bu_k = \begin{bmatrix} 1 & 0 & \Delta t & 0 \\ 0 & 1 & 0 & \Delta t \\ 0 & 0 & 1 & 0 \\ 0 & 0 & 0 & 1 \end{bmatrix} \cdot \begin{bmatrix} x_{k-1} \\ y_{k-1} \\ \dot{x}_{k-1} \\ \dot{y}_{k-1} \end{bmatrix} + \begin{bmatrix} \frac{(\Delta t)^2 a_x}{2} \\ \frac{(\Delta t)^2 a_y}{2} \\ \Delta t \, a_x \\ \Delta t \, a_y \end{bmatrix}$$

Exercise 3: Multiply out the above equation and show the state vector X_k as a 4×1 matrix.

Solution 3:

$$X_k = \begin{bmatrix} x_{k-1} + \Delta t \dot{x}_{k-1} + \frac{(\Delta t)^2 a_x}{2} \\ y_{k-1} + \Delta t \dot{y}_{k-1} + \frac{(\Delta t)^2 a_y}{2} \\ \dot{x}_{k-1} + \Delta t \, a_x \\ \dot{y}_{k-1} + \Delta t \, a_y \end{bmatrix}$$

3.4.3 Objects Moving in Three-Dimensional Space

The three-dimensional case describes the states of objects like satellites, aircrafts and drones. Drones move normally in three-dimensional space. This requires inclusion of the third dimension (z-dimension). Thus, the state matrix contains six elements consisting of positions and velocities in the three dimensions. It is

$$X = \begin{bmatrix} x \\ y \\ z \\ \dot{x} \\ \dot{y} \\ \dot{z} \end{bmatrix} \tag{3.52}$$

The way the state X is expressed also determines how variables in the state matrix A are arranged. In general, its equation of motion is given by the expression:

$$X = X_0 + \dot{X} \cdot t + \frac{1}{2}\ddot{X} \cdot t^2 \tag{3.53}$$

X is a three-dimensional vector with initial position X_0 and corrections to position with corresponding velocities $\dot{X}$ and accelerations $\ddot{X}$. The first terms of the equation of motion are used as part of the term AX_{k-1} in the state equation. The acceleration part of the motion goes into the term Bu_k and the process noise is w_k.

The state matrix for the three-dimensional case should provide for updating of the state variables with velocities while the control matrix allows for updating of the state variables as a result of the acceleration of the object. That means that the 6×6 state matrix can be composed in line with the way the state vector is written here as follows:

$$AX_{k-1} = \begin{bmatrix} 1 & 0 & 0 & \Delta t & 0 & 0 \\ 0 & 1 & 0 & 0 & \Delta t & 0 \\ 0 & 0 & 1 & 0 & 0 & \Delta t \\ 0 & 0 & 0 & 1 & 0 & 0 \\ 0 & 0 & 0 & 0 & 1 & 0 \\ 0 & 0 & 0 & 0 & 0 & 1 \end{bmatrix} \begin{bmatrix} x_{k-1} \\ y_{k-1} \\ z_{k-1} \\ \dot{x}_{k-1} \\ \dot{y}_{k-1} \\ \dot{z}_{k-1} \end{bmatrix} \tag{3.54}$$

The control matrix component of the state equation has B as a 6×3 matrix leading to the product

$$Bu_k = \begin{bmatrix} \frac{(\Delta t)^2}{2} & 0 & 0 \\ 0 & \frac{(\Delta t)^2}{2} & 0 \\ 0 & 0 & \frac{(\Delta t)^2}{2} \\ \Delta t & 0 & 0 \\ 0 & \Delta t & 0 \\ 0 & 0 & \Delta t \end{bmatrix} \begin{bmatrix} a_x \\ a_y \\ a_z \end{bmatrix} \tag{3.55}$$

By considering a zero process noise, the state equation is

$$X_k = AX_{k-1} + Bu_k = \begin{bmatrix} 1 & 0 & 0 & \Delta t & 0 & 0 \\ 0 & 1 & 0 & 0 & \Delta t & 0 \\ 0 & 0 & 1 & 0 & 0 & \Delta t \\ 0 & 0 & 0 & 1 & 0 & 0 \\ 0 & 0 & 0 & 0 & 1 & 0 \\ 0 & 0 & 0 & 0 & 0 & 1 \end{bmatrix} \begin{bmatrix} x_{k-1} \\ y_{k-1} \\ z_{k-1} \\ \dot{x}_{k-1} \\ \dot{y}_{k-1} \\ \dot{z}_{k-1} \end{bmatrix}$$

$$+ \begin{bmatrix} \frac{(\Delta t)^2}{2} & 0 & 0 \\ 0 & \frac{(\Delta t)^2}{2} & 0 \\ 0 & 0 & \frac{(\Delta t)^2}{2} \\ \Delta t & 0 & 0 \\ 0 & \Delta t & 0 \\ 0 & 0 & \Delta t \end{bmatrix} \begin{bmatrix} a_x \\ a_y \\ a_z \end{bmatrix} \tag{3.56}$$

Exercise 4: Using the above equation with $\Delta t = 1$ second

(i) What is the state vector X_k?

(ii) Repeat (i) when $a_x = a_z = 0$; $a_y = -9.8$ m/sec

Notice that the matrix B is a stacking of two 3×3 identity matrices where each identity matrix is multiplied by a constant as

$$B = \begin{bmatrix} \frac{(\Delta t)^2}{2} I_3 \\ \Delta t\, I_3 \end{bmatrix} = \begin{bmatrix} \frac{(\Delta t)^2}{2} \\ \Delta t \end{bmatrix} [I_3]$$

I_3 is a 3×3 identity matrix.

3.5 Kalman Filter Models with Noise

Normally both the predicted and measurement values in Kalman filters have errors or noise. The prediction error is sometimes also referred to as process noise. The measurement error also occurs due to deficiencies in the sensor or measuring equipment. These two noise sources are accounted for in the general Kalman filter models using covariance matrices. In this section, the words noise and error will be used to mean the same thing. The following matrices are therefore defined in this section.

- P is the state covariance matrix. It is the error in prediction or estimate
- Q is the process noise covariance matrix
- R is the measurement error covariance matrix
- K is the Kalman gain

These matrices change at each time epoch and therefore will be subscripted in the analysis to follow. In general, the following relationships hold

$$\left. \begin{array}{l} P_k = A P_{k-1} A^T + Q \\[2mm] K_k = \dfrac{P_k H^T}{H P_k H^T + R} \end{array} \right\} \tag{3.57}$$

P_k is the updated error in the prediction and K_k is the updated Kalman gain. Observe from these equations, that the Kalman gain K is a ratio of the prediction error to the sum of the prediction error and measurement noise R. H is a transformation matrix. The Kalman gain is nearly one when the measurement error is nearly zero.

In Chapter 4, we will develop this concept further by showing how to calculate the P, Q and R matrices and then the Kalman gain.

References

[1] Michel van Biezen, The Kalman Filter – The Multidimensional Model.

[2] Ramsey Faragher, "Understanding the Basis of the Kalman Filter Via a Simple and Intuitive Derivation", IEEE Signal Processing Magazine, 2012, pp. 128–132.

4

Kalman Filter II

4.1 Introduction

In the treatment of Kalman filters in Chapter 3, we assumed that the process noise is zero. This is not always the case in tracking. The rain drops in our examples in Chapter 3 could be experiencing wind in the environment, which could change their velocities along the dimensions. A flying aircraft experiences turbulence thereby introducing noise into the system so that the value of altitude reported is in error. Kalman filters are developed for such situations where the theoretical system equations are no more correct due to noise in the system. To avoid confusion in subscripts and difficulties in learning Kalman filtering, double subscripting of variables between previous state and current state is avoided. Subscripting of variables is limited to single subscripts. It is hoped that by doing so, it will become easier to gain traction in learning.

4.2 Processing Steps in Kalman Filter

The processing steps in KF algorithm were introduced in Chapter 3. The same diagram is retained in Figure 4.1 in this chapter.

4.2.1 Covariance Matrices

Covariance matrices are popular in statistical data analytic applications. They are used to compare statistical data to establish the levels of similarity between them. In this section, how to compute covariance and covariance matrices are introduced first, followed by their use in Kalman filtering. The following definitions apply for the rest of the discussions in this chapter. Let

x_i be individual measurements

$\bar{X}$ be the mean or average of the measurements;

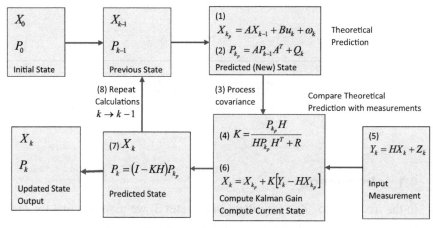

Figure 4.1   Processing steps for kalman filters.

$(x_i - \bar{X})$ be the deviation from the mean so that it takes either positive or negative values, which depend on the difference and

$(x_i - \bar{X})^2$ be the square of the deviation and is a positive number;

The variance of the data sequence $\{X\}$ of N samples is

$$\sigma_x^2 = \frac{\sum\limits_{i=1}^{N}(x_i - \bar{X})^2}{N} \tag{4.1}$$

Standard deviation is the square root of variance and is given by the equation

$$\sigma_x = \sqrt{\frac{\sum\limits_{i=1}^{N}(x_i - \bar{X})^2}{N}} \tag{4.2}$$

The variance is a measure of how the data sequence lies around the mean value. The variance is a positive number. Suppose there is a second data sequence $\{Y\}$ of mean $\bar{Y}$ also of the same length at the sequence $\{X\}$. The covariance of the two data sequences is

$$\sigma_x\sigma_y = \frac{\sum\limits_{i=1}^{N}(x_i - \bar{X})(y_i - \bar{Y})}{N} \tag{4.3}$$

The covariance is the product of the two standard deviations for X and Y data sequences. With the above standard definitions, the covariance matrices for one-, two- and three-dimensional Kalman filter are defined by the following expressions:

One Dimension (1D)

$$\rho = \sigma_x^2 = \left[\frac{\sum\limits_{i=1}^{N}(x_i - \bar{X})^2}{N} \right] \tag{4.4}$$

The covariance is a measure of the energy in the signal and is useful in applications where the signal-to-noise ratio of a signal needs to be determined.

Two Dimensions (2D)

For two dimensions, the covariance matrix is a 2×2 matrix given by the expression

$$\rho = \begin{bmatrix} \sigma_x^2 & \sigma_x\sigma_y \\ \sigma_x\sigma_y & \sigma_y^2 \end{bmatrix} = \left[\begin{array}{cc} \frac{\sum\limits_{i=1}^{N}(x_i-\bar{X})^2}{N} & \frac{\sum\limits_{i=1}^{N}(x_i-\bar{X})(y_i-\bar{Y})}{N} \\ \frac{\sum\limits_{i=1}^{N}(y_i-\bar{Y})(x_i-\bar{X})}{N} & \frac{\sum\limits_{i=1}^{N}(y_i-\bar{Y})^2}{N} \end{array} \right] \tag{4.5}$$

Exercise

The following data represent readings from a temperature and humidity sensors over a period of several time instants. Computer their means and variances.

Temperature	Humidity
26.1349	58.9679
26.1344	58.9679
26.1340	58.9679
26.1336	58.9678
26.1333	58.9678
26.1330	58.9678
26.1327	58.9678
26.1326	58.9678
26.1333	58.9677
26.1339	58.9677

Three Dimensions (3D)

$$\rho = \begin{bmatrix} \sigma_x^2 & \sigma_x\sigma_y & \sigma_x\sigma_z \\ \sigma_y\sigma_x & \sigma_y^2 & \sigma_y\sigma_z \\ \sigma_z\sigma_x & \sigma_z\sigma_y & \sigma_z^2 \end{bmatrix}$$

$$= \begin{bmatrix} \dfrac{\sum\limits_{i=1}^{N}(x_i-\bar{X})^2}{N} & \dfrac{\sum\limits_{i=1}^{N}(x_i-\bar{X})(y_i-\bar{Y})}{N} & \dfrac{\sum\limits_{i=1}^{N}(x_i-\bar{X})(z_i-\bar{Z})}{N} \\ \dfrac{\sum\limits_{i=1}^{N}(y_i-\bar{Y})(x_i-\bar{X})}{N} & \dfrac{\sum\limits_{i=1}^{N}(y_i-\bar{Y})^2}{N} & \dfrac{\sum\limits_{i=1}^{N}(y_i-\bar{Y})(z_i-\bar{Z})}{N} \\ \dfrac{\sum\limits_{i=1}^{N}(z_i-\bar{Z})(x_i-\bar{X})}{N} & \dfrac{\sum\limits_{i=1}^{N}(z_i-\bar{Z})(y_i-\bar{Y})}{N} & \dfrac{\sum\limits_{i=1}^{N}(z_i-\bar{Z})^2}{N} \end{bmatrix} \qquad (4.6)$$

The standard deviation of the data provides a means of assessing the nature of the distribution of a data sequence. Normally, about 68% of all the measurements lie within ($\pm\sigma$) one standard deviation from the mean. All measurement samples also lie within $\pm\sigma^2$ from the mean value.

4.2.2 Computation Methods for Covariance Matrix

4.2.2.1 Manual method

Several methods exist for computing the covariance matrix including manual and deviation matrix methods. The manual method is tedious and prone to errors. The deviation matrix method is more tractable and amenable for software or programming approach. To illustrate the manual approach, the following exercise is given.

Exercise

Three sensors were deployed in an environment to measure the concentration of ethanol in the environment with the following results:

S1_max	S2_max	S3_max
0.09353	0.09985	0.08644
0.08080	0.08781	0.14446
0.10897	0.10989	0.16347
0.11300	0.10598	0.09914
0.15199	0.15098	0.20885
0.14486	0.14459	0.13876

(a) Compute the mean readings from each gas detection sensor
(b) Calculate the variance and standard deviations for each of the gas detection sensor
(c) What is the covariance matrix P for the three ethanol sensors?
(d) Show that all the samples for each data sequence lie within one variance of that data sequence.

Example

In this section, a numerical example is provided. The following table contains readings from three sensors S7, S8 and S9. The covariance of each record is computed

S7	S8	S9
0.09442	0.29945	0.23588
0.09639	0.22321	0.14178
0.11959	0.28804	0.21071
0.11036	0.30146	0.26484
0.1665	0.34434	0.23755
0.15396	0.30497	0.23874

Solution

$$S\bar{7} = \frac{\sum_{i=1}^{6}(S\bar{7}_i)}{6}$$

$$= \frac{1}{6}[0.09442 + 0.09639 + 0.11959 + 0.11036 + 0.1665 + 0.15396]$$

$$= 0.12353$$

$$\sigma_{S7}^2 = \frac{\sum_{i=1}^{6}(S7_i - S\bar{7})^2}{6}$$

$$= \frac{1}{6}\left[(0.09442 - 0.12353)^2 + (0.09639 - 0.12353)^2\right.$$

$$+ (0.11959 - 0.12353)^2 + (0.11036 - 0.12353)^2$$

$$+ (0.1665 - 0.12353)^2 + (0.15396 - 0.12353)^2\Big]$$

$$= 0.000745$$

$$\sigma_{S7} = \sqrt{\frac{\sum\limits_{i=1}^{N}(S7_i - S\bar{7})^2}{N}} = 0.0273$$

$$S\bar{8} = \frac{\sum\limits_{i=1}^{6}(S8_i)}{6}$$

$$= \frac{1}{6}[0.29945 + 0.22321 + 0.28804 + 0.30146 + 0.34434 + 0.30497]$$

$$= 0.29358$$

$$\sigma_{S8}^2 = \frac{\sum\limits_{i=1}^{6}(S8_i - S\bar{8})^2}{6}$$

$$= \frac{1}{6}\left[(0.29945 - 0.29358)^2 + (0.22321 - 0.29358)^2\right.$$

$$+ (0.28804 - 0.29358)^2 + (0.30146 - 0.29358)^2$$

$$\left. + (0.34434 - 0.29358)^2 + (0.30497 - 0.29358)^2\right]$$

$$= 0.001298$$

$$\sigma_{S8} = \sqrt{\frac{\sum\limits_{i=1}^{6}(S8_i - S\bar{8})^2}{N}}$$

$$= 0.03602$$

$$S\bar{9} = \frac{\sum\limits_{i=1}^{6}(S9_i)}{6}$$

$$= \frac{1}{6}[0.23588 + 0.14178 + 0.21071 + 0.26484 + 0.23755 + 0.23874]$$

$$= 0.22158$$

$$\sigma_{S9}^2 = \frac{\sum\limits_{i=1}^{6}(S9_i - S\bar{9})^2}{6}$$

$$= \frac{1}{6}\left[(0.23588 - 0.22158)^2 + (0.21071 - 0.22158)^2\right.$$

$$+ (0.14178 - 0.22158)^2 + (0.26484 - 0.22158)^2$$
$$+ (0.23755 - 0.22158)^2 + (0.23874 - 0.22158)^2]$$
$$= 0.0015186$$

$$\sigma_{S9} = \sqrt{\frac{\sum_{i=1}^{6}(S9_i - \bar{S9})^2}{N}} = 0.039$$

The covariance matrix P is

$$P = \begin{bmatrix} \sigma_{S7}^2 & \sigma_{S7}\sigma_{S8} & \sigma_{S7}\sigma_{S9} \\ \sigma_{s8}\sigma_{s7} & \sigma_{s8}^2 & \sigma_{s8}\sigma_{s9} \\ \sigma_{s9}\sigma_{s7} & \sigma_{s9}\sigma_{s8} & \sigma_{s9}^2 \end{bmatrix}$$

$$= \begin{bmatrix} 0.000745 & 0.000983 & 0.0010647 \\ 0.000983 & 0.001298 & 0.00140478 \\ 0.0010647 & 0.00140478 & 0.0015186 \end{bmatrix}$$

4.2.2.2 Deviation matrix computation method

One of the easiest ways of computing covariances suitable for programming is the use of deviation matrices. The deviation of a matrix A is defined by the expression

$$\alpha = A - \frac{[I]A}{N} \tag{4.7}$$

where N is the number of values in the column of matrix A and I is the identity matrix of equal size with A. The covariance of the matrix A is then given by the product of deviation matrix and its transpose as

$$\sigma = \alpha^T \alpha \tag{4.8}$$

Example

Consider the prices of three stock portfolios, B, C and G each having three stocks.

$$P = \begin{bmatrix} 80 & 40 & 90 \\ 90 & 70 & 60 \\ 70 & 70 & 60 \end{bmatrix}.$$

Find the covariance matrix for the three portfolios.

Solution

The deviation matrix is

$$
\alpha = \begin{bmatrix} 80 & 40 & 90 \\ 90 & 70 & 60 \\ 70 & 70 & 60 \end{bmatrix} - \begin{bmatrix} 1 & 1 & 1 \\ 1 & 1 & 1 \\ 1 & 1 & 1 \end{bmatrix} \times \begin{bmatrix} 80 & 40 & 90 \\ 90 & 70 & 60 \\ 70 & 70 & 60 \end{bmatrix} \frac{1}{3}
$$

$$
= \begin{bmatrix} 0 & -20 & 20 \\ 10 & 10 & -10 \\ -10 & 10 & -10 \end{bmatrix}
$$

$$
\alpha^T \alpha = \begin{bmatrix} 0 & 10 & -10 \\ -20 & 10 & 10 \\ 20 & -10 & -10 \end{bmatrix} \times \begin{bmatrix} 0 & -20 & 20 \\ 10 & 10 & -10 \\ -10 & 10 & -10 \end{bmatrix}
$$

$$
= \begin{bmatrix} 200 & 0 & 0 \\ 0 & 600 & -600 \\ 0 & -600 & 600 \end{bmatrix}
$$

Therefore, the covariance is:

$$
\rho = \alpha^T \alpha = \begin{bmatrix} \sigma_B^2 & \sigma_B\sigma_C & \sigma_B\sigma_G \\ \sigma_C\sigma_B & \sigma_C^2 & \sigma_C\sigma_G \\ \sigma_G\sigma_B & \sigma_G\sigma_C & \sigma_G^2 \end{bmatrix} = \begin{bmatrix} 200 & 0 & 0 \\ 0 & 600 & -600 \\ 0 & -600 & 600 \end{bmatrix}
$$

$$
\sigma_B = 10\sqrt{2}; \quad \sigma_C = 10\sqrt{6}; \quad \sigma_G = -10\sqrt{6}
$$

Four of the covariance matrix values are zero. In other words, there is no dependence between stock B and stock C and G. Also, there is no dependence between stocks B and G. Those error variables are independent of each other. Also, there is negative covariance between stock C and G, which means when one goes up in price, the other goes down.

The covariance matrices are highly informative. Entries in the matrices are products of standard deviations that allows each entry to be composed as products of two or more standard deviations. Thus, for example, a process of the covariance matrix can be created rather quickly for a process of

N (columns) components such as

$$\rho = \begin{bmatrix} \sigma_{11}^2 & \sigma_1\sigma_{j+1} & \sigma_1\sigma_{j+2} & \cdots & \sigma_1\sigma_{j+N-1} \\ \sigma_2\sigma_1 & \sigma_{22}^2 & \sigma_2\sigma_{j+2} & \cdots & \sigma_2\sigma_{j+N-1} \\ \sigma_3\sigma_1 & \sigma_3\sigma_2 & \sigma_{33}^2 & \cdots & \sigma_3\sigma_{j+N-1} \\ \vdots & \vdots & \vdots & \ddots & \vdots \\ \sigma_N\sigma_1 & \sigma_N\sigma_2 & \sigma_N\sigma_3 & \cdots & \sigma_{NN}^2 \end{bmatrix}$$

Example

A rain droplet at a height of 3000 m is falling with a velocity of 300 m/s and acceleration 9.8 m/s^2. If the velocity has standard deviation 1.2 m/s and the height has a deviation of 5.3 m, calculate the state covariance matrix for the process.

Solution

Given: $\sigma_{\dot{x}} = 3.5 \ m/s^2; \sigma_x = 5.3 \ m$.

The state of the process is $X = \begin{bmatrix} x \\ \dot{x} \end{bmatrix}$

The state transition matrix is given as follows:

$$P = \begin{bmatrix} \sigma_x^2 & \sigma_x\sigma_{\dot{x}} \\ \sigma_{\dot{x}}\sigma_x & \sigma_{\dot{x}}^2 \end{bmatrix} = \begin{bmatrix} (3.5)^2 & (3.5) \times (5.3) \\ (5.3) \times (3.5) & (5.3)^2 \end{bmatrix}$$

$$= \begin{bmatrix} 12.25 & 18.55 \\ 18.55 & 28.09 \end{bmatrix}$$

4.2.3 Iterations in Kalman Filter

In this section, we will look at tracking of a military helicopter flying at a speed of 220 m/s at a height of 4000 m initially. The initial state and measurements are as provided. The objective of this section is to use the given data to provide a realistic example of how Kalman filter works.

$$X_0 = \begin{bmatrix} x_0 = 4000 \\ v_{0x} = 240 \end{bmatrix};$$

$$y_0 = 5000 \ \text{m}; v_{0y} = 120 \ \text{m/sec}$$

Starting conditions

Acceleration: $a_{xo} = 3$ m/s^2; velocity: $v_{xo} = 3$ m/sec; $\Delta x = 30$ m and $\Delta t = 1$ sec

Process errors: $\Delta P_x = 25$ m; $\Delta P_{\dot{X}} = 5$ m/sec;
Error in observations: $\Delta x = 24$ m; $\Delta \dot{x} = 4$ m/sec;

Observation Vector:

X	Value (m)	v_x	Value (m/s)
X_0	4000	v_{0x}	240
X_1	4220	v_{1x}	235
X_2	4430	v_{2x}	245
X_3	4650	v_{3x}	242
X_4	4810	v_{4x}	250

Figure 4.1 provides the algorithmic steps in Kalman filtering. In this section, we describe the eight steps. All the error matrices are for simplicity set to zero. The control variable is however used in the processing. The eight (8) steps are as follows:

(1) State Prediction

From Figure 4.1, the state equation is given as

$$X_{k_p} = AX_{k-1} + Bu_k + \omega_k \tag{4.9}$$

Since $\omega_k = 0$, the state prediction equation reduces to

$$X_{k_p} = AX_{k-1} + Bu_k + \omega_k = 0 \tag{4.10}$$

The state dynamic equations covering the motion of an aircraft were given in Chapter 3 and is

$$x = x_0 + vt + at^2 = x_0 + \dot{x}t + \frac{1}{2}\ddot{x}t^2 \tag{4.11}$$

$$Y_k = CX_k + z_k \tag{4.12}$$

where
  A is the state matrix
  B is the control matrix
  C is the measurement matrix
  u is the control variable
  w is the process noise

k is the time index and for each iteration k advances by 1 time slot Δt

z_k is the measurement noise and for this section is set to zero

The state matrix A and the control matrix were also given in Chapter 3 as

$$A = \begin{bmatrix} 1 & \Delta t \\ 0 & 1 \end{bmatrix} \quad \text{and} \quad B = \begin{bmatrix} 0.5\Delta t^2 \\ \Delta t \end{bmatrix}$$

With these matrices and data given, we can now undertake tracking of the helicopter as

$$X_{k_p} = \begin{bmatrix} 1 & \Delta t \\ 0 & 1 \end{bmatrix} X_{k-1} + \begin{bmatrix} 0.5\Delta t^2 \\ \Delta t \end{bmatrix} u_k + 0$$

$$= \begin{bmatrix} 1 & 1 \\ 0 & 1 \end{bmatrix} \times \begin{bmatrix} x_0 \\ v_{x0} \end{bmatrix} + \begin{bmatrix} 0.5 \\ 1 \end{bmatrix} \times [a_{x0}]$$

Therefore, based on our data

$$X_{k_p} = \begin{bmatrix} 1 & 1 \\ 0 & 1 \end{bmatrix} \times \begin{bmatrix} 4000 \\ 240 \end{bmatrix} + \begin{bmatrix} 0.5 \\ 1 \end{bmatrix} \times [3] = \begin{bmatrix} 4240 \\ 240 \end{bmatrix} + \begin{bmatrix} 1.5 \\ 3 \end{bmatrix}$$

$$X_{k_p} = \begin{bmatrix} 4241.5 \\ 243 \end{bmatrix}$$

Note the predicted state has corrections in position due to velocity and acceleration and corrections in velocity as a result of the acceleration through the B matrix.

(2) Process Covariance Matrix

This step for computing the process covariance matrix is done once at the beginning and it is updated in subsequent steps. Process covariance matrix is determined by the equation

$$P_{k-1} = \begin{bmatrix} (\Delta x)^2 & \Delta x \Delta v_x \\ \Delta v_x \Delta x & (\Delta v_x)^2 \end{bmatrix} \tag{4.13}$$

Initial values for these state variables were given earlier and are inserted into the equation to give:

$$P_{k-1} = \begin{bmatrix} (\Delta x)^2 & \Delta x \Delta v_x \\ \Delta v_x \Delta x & (\Delta v_x)^2 \end{bmatrix} = \begin{bmatrix} (30)^2 & 30 \times 3 \\ 3 \times 30 & (3)^2 \end{bmatrix}$$

$$= \begin{bmatrix} 900 & 90 \\ 90 & 9 \end{bmatrix}$$

(3) Predicted Covariance Matrix

With the value of the state covariance matrix obtained from round (2) for the first time, we can now predict the state covariance matrix to use in subsequent calculations. The expression to use in creating the predicted covariance matrix is given in Figure 4.1 and is:

$$P_{k_p} = AP_{k-1}A^T + Q_k \tag{4.14}$$

We are given that the prediction error is initially zero. The predicted covariance matrix becomes

$$P_{k_p} = AP_{k-1}A^T + (Q_k = 0)$$

$$= \begin{bmatrix} 1 & 1 \\ 0 & 1 \end{bmatrix} \times \begin{bmatrix} 900 & 0 \\ 0 & 9 \end{bmatrix} \times \begin{bmatrix} 1 & 0 \\ 1 & 1 \end{bmatrix} + 0$$

$$= \begin{bmatrix} 900 & 9 \\ 0 & 9 \end{bmatrix} \times \begin{bmatrix} 1 & 0 \\ 1 & 1 \end{bmatrix} = \begin{bmatrix} 909 & 9 \\ 9 & 9 \end{bmatrix}$$

Again we set the cross-covariance terms to zero so that we have

$$P_{k_p} = \begin{bmatrix} 909 & 0 \\ 0 & 9 \end{bmatrix}$$

We can simplify this computation if we, at this time, neglect the cross-covariance terms in the A matrix. This is because they do not really affect the distance and velocity covariance values. If we set them to zero initially, we can speed up the computation of the predicted covariance matrix. We have chosen not to eliminate them to help the reader go through the overall computation of the predicted covariance matrix.

(4) Kalman Gain

The Kalman gain is a term when the state vector needs to be updated. The Kalman gain expression is given as

$$K = \frac{P_{k_p}H}{HP_{k_p}H^T + R} \tag{4.15}$$

How do we determine the H matrix? H is called the observation matrix. It converts the predicted covariance matrix into the correct form. To calculate the Kalman gain, the observation covariance error matrix is required. It is therefore of the same form as the A matrix. In our case, given the observation errors $\Delta x = 24$ m; $\Delta \dot{x} = 4$ m/sec, the error covariance matrix is:

$$R = \begin{bmatrix} (\Delta x)^2 & \Delta x \Delta v_x \\ \Delta v_x \Delta x & (\Delta v_x)^2 \end{bmatrix} = \begin{bmatrix} (24)^2 & 24 \times 4 \\ 4 \times 24 & (4)^2 \end{bmatrix}$$

$$= \begin{bmatrix} 576 & 96 \\ 96 & 16 \end{bmatrix}$$

With this matrix in hand, it is now possible to calculate the Kalman gain by substituting the relevant matrices. The H matrix is an identity matrix. Therefore:

$$K = \frac{P_{k_p} H}{H P_{k_p} H^T + R}$$

$$= \frac{\begin{bmatrix} 909 & 0 \\ 0 & 9 \end{bmatrix} \times \begin{bmatrix} 1 & 0 \\ 0 & 1 \end{bmatrix}}{\begin{bmatrix} 1 & 0 \\ 0 & 1 \end{bmatrix} \times \begin{bmatrix} 909 & 0 \\ 0 & 9 \end{bmatrix} \times \begin{bmatrix} 1 & 0 \\ 0 & 1 \end{bmatrix} + \begin{bmatrix} 576 & 96 \\ 96 & 16 \end{bmatrix}}$$

$$K = \frac{\begin{bmatrix} 909 & 0 \\ 0 & 9 \end{bmatrix}}{\begin{bmatrix} 909 & 0 \\ 0 & 9 \end{bmatrix} + \begin{bmatrix} 576 & 96 \\ 96 & 16 \end{bmatrix}} = \frac{\begin{bmatrix} 909 & 0 \\ 0 & 9 \end{bmatrix}}{\begin{bmatrix} 1485 & 96 \\ 96 & 25 \end{bmatrix}}$$

$$= \begin{bmatrix} 909 & 0 \\ 0 & 9 \end{bmatrix} x \begin{bmatrix} .0008958 & -0.00344 \\ -0.00344 & .0532 \end{bmatrix}$$

$$K = \begin{bmatrix} 0.8143 & -0.03096 \\ -0.03096 & 0.4788 \end{bmatrix}$$

The inverse of the matrix in the denominator is required and may be computed using any of the methods for finding the inverse of a regular matrix. Doing so gives the Kalman gain

$$K = \begin{bmatrix} 0.8143 & -0.03096 \\ -0.03096 & 0.4788 \end{bmatrix}$$

(5) Calculating New Observation

$$Y_k = H X_k + Z_k \tag{4.16}$$

Since the sensor noise error is inherent to the device, for now it is set to zero or Z_k.

The new observation therefore becomes

$$Y_k = HX_k + (Z_k = 0)$$

$$Y_k = \begin{bmatrix} 1 & 0 \\ 0 & 1 \end{bmatrix} \times \begin{bmatrix} 4220 \\ 235 \end{bmatrix} = \begin{bmatrix} 4220 \\ 235 \end{bmatrix}$$

(6) Predicting Current State

In step (1), we calculated $X_{k_p} = \begin{bmatrix} 4241.5 \\ 243 \end{bmatrix}$

In step (4), we calculated $K = \begin{bmatrix} 0.8143 & -0.03096 \\ -0.03096 & 0.4788 \end{bmatrix}$

The current measurement is $Y_k = \begin{bmatrix} 4220 \\ 235 \end{bmatrix}$

Therefore, the updated current state is:

$$X_k = X_{k_p} + K[Y_k - HX_{k_p}] = \begin{bmatrix} 4241.5 \\ 243 \end{bmatrix} + \begin{bmatrix} 0.8143 & -0.03096 \\ -0.03096 & 0.4788 \end{bmatrix}$$

$$\times \left\{ \begin{bmatrix} 4220 \\ 235 \end{bmatrix} - \begin{bmatrix} 1 & 0 \\ 0 & 1 \end{bmatrix} \times \begin{bmatrix} 4241.5 \\ 243 \end{bmatrix} \right\}$$

$$X_k = \begin{bmatrix} 4241.5 \\ 243 \end{bmatrix} + \begin{bmatrix} 0.8143 & -0.03096 \\ -0.03096 & 0.4788 \end{bmatrix}$$

$$\times \left\{ \begin{bmatrix} -21.5 \\ -8 \end{bmatrix} \right\} = \begin{bmatrix} 4241.5 \\ 243 \end{bmatrix} + \begin{bmatrix} -17.824 \\ -3.165 \end{bmatrix}$$

$$= \begin{bmatrix} 4223.745 \\ 239.835 \end{bmatrix}$$

(7) Updating Process Covariance Matrix

From the previous steps, we have obtained the following matrices

$$K = \begin{bmatrix} 0.8143 & -0.03096 \\ -0.03096 & 0.4788 \end{bmatrix} \quad \text{and} \quad P_{k_p} = \begin{bmatrix} 909 & 0 \\ 0 & 9 \end{bmatrix}$$

Therefore, all the requirements for updating the process covariance matrix are in hand. This is done with the expression

$$P_k = (I - KH)P_{k_p}$$

$$= \left(\begin{bmatrix} 1 & 0 \\ 0 & 1 \end{bmatrix} - \begin{bmatrix} 0.8143 & -0.03096 \\ -0.03096 & 0.4788 \end{bmatrix} \times \begin{bmatrix} 1 & 0 \\ 0 & 1 \end{bmatrix} \right) \times \begin{bmatrix} 909 & 0 \\ 0 & 9 \end{bmatrix}$$

$$P_k = \begin{bmatrix} 0.1857 & 0.03096 \\ 0.03096 & 0.5212 \end{bmatrix} \times \begin{bmatrix} 909 & 0 \\ 0 & 9 \end{bmatrix}$$

$$P_k = \begin{bmatrix} 168.80 & 0.27864 \\ 28.14 & 4.6908 \end{bmatrix}$$

5

Genetic Algorithm

5.1 Introduction

Decision making and the cue to someone to take a line of action and decide what to do, what is best and what is more financially smart based on given data is an area of great interest in industry in modern data analytic businesses. Many businesses have huge data archives from which decision could be reached on issues like which product to push to the market, which market and when. In logistics and transportation, routing of traffic to minimize delays remains an area of great interest. Taking a route among many routes could result in savings of fuel, reduce costs and save time. Genetic algorithms belong to a class of search algorithms by mimicking how biological processes evolve and by extension help to unravel how natural and commercial processes adapt to changing conditions. They can be used to design software tools for decision making and for designing robust systems based on the inter-relationships between system parameters. In doing so, they belong to a class of optimization algorithms, the so-called evolutionary computing algorithms. They are a class of optimization problems. For that reason, GAs are used to find maxima or minima of functions.

5.2 Steps in Genetic Algorithm

The generic genetic algorithm consists of six steps:

(1) *Population*: all GAs rely on starting with a random population of chromosomes. They are typically sets of binary bits, which represent each member of the population.
(2) *Fitness function*: a fitness function provides the means for assessing the fitness of the next generation of the population. Therefore, it is necessary to create a fitness function which is used for optimization.

75

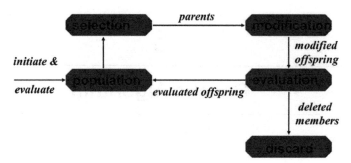

Figure 5.1   Evolution cycle.

(3) **Selection algorithm:** a selection algorithm for which chromosomes will reproduce in the next generation is based on the fittest or best solution.

(4) **Crossover:** Crossover of genes introduces randomness into the population. It is used to produce the next generation of chromosomes. In this process, offspring are selected to reproduce; crossover is a concept that is borrowed from genetics. Normally, it is really a sexual reproduction process in which two parents mate and exchange genetic materials to create superior offspring. In GA, single-point or two-point crossovers are popular.

(5) **Mutation:** in this process, a random mutation of chromosomes in the new generation is applied. Mutation introduces diversity into the population.

(6) **Survivors:** a method of selecting survivors.

These six steps will be described in detail in this chapter. Genetic algorithms depend on mimicking biological evolution of species. It relies on survival of the fittest. The fittest are selected to reproduce while the weakest are mostly ignored. In other words, the best solutions survive while bad solutions are left to die.

5.3 Terminology of Genetic Algorithms (GAs)

Biological species may be described in terms of populations and the characteristics of the populations. In this section, this model is used to describe the basic terminology of genetic algorithms.

Population: The population in a GA is defined as a subset of all possible solutions to the problem in hand. In other words, there is a bigger set of possible solutions out of which a subset of solutions is used as the population.

Chromosome: A chromosome in the sense of genetic algorithm is one of the possible solutions in the population for the given problem. Chromosomes consist of elements and their positions are called genes.

Gene: A gene is an element position within the chromosome. A gene in a chromosome takes a value and the value is called ***allele***.

Genotype: This is a population within the computation space. There is a second definition of population in the actual real-world solution space called phenotype.

Phenotype: From the above definition, we therefore define a phenotype as the population in the actual real-world solution space. In practice, a mathematical transformation function is used to provide the connection between the genotype and phenotype. This transformation is called decoding.

Encoding: A GA normally handles generations of 'biological organisms', which are represented with chromosomes C. To model the GA, the chromosomes need to be fully and efficiently determined to ensure that during each generation, the iteration does not descend into a locked position to hinder variation. Each chromosome is described mathematically with binary numbers or bits.

Encoding is the transformation of the parameters of the problem in hand into a chromosome. Consider an N parameter problem with parameter set designated as $p_i \in (1 \leq i \leq N)$. The chromosome is the sequence of values

$$C = [p_1, \ldots, p_N] \tag{5.1}$$

The binary representation of all the p_i are concatenated to form the chromosome. The choice of parameters is a design exercise in the use of GA.

Decoding: This is a transformation between the phenotype and genotype spaces. These GA terms are illustrated in Figure 5.2.

As shown in Figure 5.2, the population consists of individuals. An individual is a set of chromosomes with genes. Inside each gene is an allele. The table shows a population of 10 distinct entities of different fitness attributes, which we are yet to define. The next section describes how fitness functions are modelled.

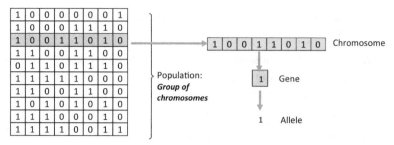

Figure 5.2 Terminology of genetic algorithms.

5.4 Fitness Function

Fitness functions are also called evaluation functions. In genetic algorithms, they are used for ranking of populations as functions of the fitness of population members. In other words, a fitness function assesses how close to optimal solution a given solution is. The fitness function is normally the function to be optimized. Each solution of the GA can therefore be ranked. This leads to selection and the survival of the fitness in the population. Solutions to problems are usually groups of chromosomes, which makes it difficult to rank them without a criterion that is fair on all possible solutions. A fitness function therefore is used to give a score to each solution and to rank them.

How therefore should fitness functions be defined or created? Several criteria have been suggested, which include the following.

5.4.1 Generic Requirements of a Fitness Function

Suitable fitness functions should satisfy the following requirements:

(1) First, a fitness function should be good enough to provide results that are intuitive to the problem. Thus, it should lead to a clear distinction between the best and worst solutions.
(2) Second, it should measure in a quantitative manner how fit or good a given solution is in terms of solving the problem in hand.
(3) Third, a fitness function should be easily and efficiently implementable. It therefore should not pose a bottleneck to solving the problem.
(4) Fourth, the fitness function should have a clear definition and not fuzzy in nature and easily understandable to the user in assessing how fitness scores are calculated with it.
(5) The degree to which a fitness function is able to discriminate between members of the population is important. It should be able to distinguish between members of the population.

(6) Members of the population who share similar characteristics should also have similar fitness functions.

(7) It should point towards the solution to the problem. How quickly this is achieved is essential.

Example 1: Finding the best three stocks to buy to maximize profit within a set of options.

Let the three stocks have yield values x, y and z. The problem in hand is to find the best set of returns to maximize profit p. This means that the objective function to maximize is

$$p = x + y + z \qquad (5.2)$$

We must therefore minimize loss in profit or deviation from $p|p - (x + y + z)|$ and reduce the loss to as close to zero as possible. Therefore, when the loss is very small, the inverse of the loss should be very high. We may therefore use the fitness function

$$f = 1/|p - (x + y + z)| \qquad (5.3)$$

Example 2: To illustrate how GA works, consider the following example. A group of students were chosen at random to represent a University at a technical competition. The selectors are not yet sure if they have made the right choice and wants to run an algorithm to pick the best set of students to represent the university. The selectors thought it out that they will base their choice on how they performed in 10 subjects. Since there are many students who satisfy the criteria, it decided to use GA for the selection of the best fit students.

They chose 6 initial students as the starting population. Whenever a student passes a subject, a 1 is recorded, and whenever the student fails a subject, a zero is recorded. Students with the highest sum of ones are deemed to be the fittest initially. This is irrespective of the subject. Are they correct? They chose to use GA to maximize their decision.

Population	Sum of 1s	Fitness
$s_1 = 1111010101$	$f(s_1) = 7$	7/34
$s_2 = 0111000101$	$f(s_2) = 5$	5/34
$s_3 = 1110110101$	$f(s_3) = 7$	7/34
$s_4 = 0100010011$	$f(s_4) = 4$	4/34
$s_5 = 1110111101$	$f(s_5) = 8$	8/34
$s_6 = 0100110000$	$f(s_6) = 3$	3/34

Fitness Function: Let the fitness function be the number of 1s a student has as ratio of the total 1s in the population (i.e., a probability function). The fitness function is

$$p(i) = \frac{f(s_i)}{\sum_{i=1}^{6} f(s_i)} = \frac{f(s_i)}{34} \tag{5.4}$$

This is given as the array

Fitness
7/34
5/34
7/34
4/34
8/34
3/34

The fitness function gives probabilities and the individuals are ranked in terms of their probabilities and in the figure given here from the highest probable to the lowest probable individuals. This is shown in Figure 5.3. As shown in Figure 5.3, the fitness function takes an input and provides a normalized output. The normalized input is used for ranking the population.

Fitness for an individual in the population is defined by a relationship of the type:

$$f = \frac{f_i}{\bar{f}} \tag{5.5}$$

where f_i is an evaluation associated with the individual i in the population and $\bar{f}$ is an average evaluation associated with all the individuals in the population. Typically, Equation (5.4) can be represented as a probability function. For example, the fitness of the kth string x^k is $f_k = g(x^{(k)})$, which is given to each chromosome in the population. Having a high fitness is a sign of better fitness and low fitness is a less desirable attribute for selection of a chromosome or member of the population.

A good fitness function should be able to classify a population efficiently. It should also have lower computation complexity.

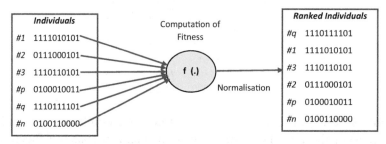

Figure 5.3   Computation of fitness function and ranking of individuals.

5.5 Selection

In each generation, a set from the existing population is selected to breed. Selection is based on the fitness function defined by the GA programmer. In one form, individuals are rated using the fitness function on each solution. A random selection may also be used, but the method could take a longer time to go through the generations. Usually it is prudent to also include the less fit solutions in the selection. This introduces diversity into the population and prevents premature convergence of the algorithm. Two selection methods are popular and include the roulette wheel and tournament selection methods.

5.5.1 The Roulette Wheel

The roulette wheel selection method is a probabilistic approach. Individuals in the population are assigned probabilities proportional to their fitness, as a ratio of the total fitness of the population. Based on the probabilities, two individuals are chosen in a random manner by spinning the roulette wheel and where it lands, the individual at the stop is chosen. The roulette wheel is given a second spin, and where it lands, a second person is chosen to reproduce with the first chosen individual.

The pseudo-code for the roulette wheel would look like this

```
for all K members of the population {
        sum += fitness of individual k;
        %sum all the fitness for the individuals
}
for all K members of the population {
        probability = sum of probabilities +
        (fitness k / sum over all K);
        sum of probabilities += probability;
}
loop until the new population is full {
        do this two times
                number p = Random between 0 and 1
                for all members of the population {
                        if number p > probability, but is
                        less than the next probability
                        then the person has been selected
                }
        }
Create offspring here;
```

5.5.2 Crossover

The next state of the GA computation is crossover. At this stage, chromosomes from two parents are shared through the process of crossover. Crossover operations are described in this section first with expressions and then an example. Crossover may occur at only one position or many positions in the chromosome.

5.5.2.1 Single-position crossover

In the simple one-position crossover, the two chromosomes are cut at identical points. Each of the parts are transferred to each other as in Figure 5.4. In other words, for the chromosomes

$$x = \langle x_1, \quad x_2, \ldots, x_n \rangle$$
$$y = \langle y_1, \quad y_2, \ldots, y_n \rangle \tag{5.6a}$$

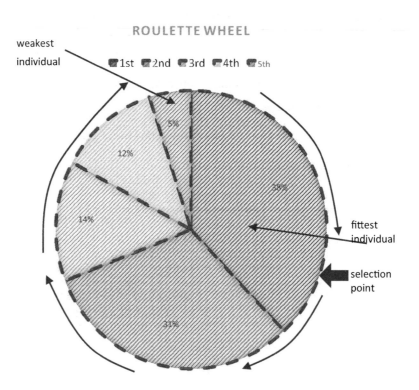

Figure 5.4   Roulette wheel.

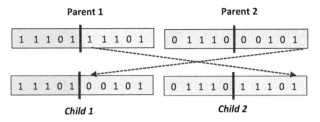

Figure 5.5 Crossover.

The index i for each allele in the chromosome $i \in N_{n-1}$ is chosen to be the position of crossover, the crossover operation performs the permutations:

$$x' = \langle x_1, \ldots, x_i, \quad y_{i+1}, \ldots, y_n \rangle$$
$$y' = \langle y_1, \ldots, y_i, \quad x_{i+1}, \ldots, x_n \rangle \tag{5.6b}$$

The two chromosomes involved in a crossover operation are called mates. The most common type of crossover is a single crossover, a common point within the chromosome. Some of the alleles from two parents are crossed over to create new offspring. In the binary example, consider the chromosomes given in Figure 5.5. The mates exchange alleles starting from given positions.

Figure 5.5 shows crossover between Parent 1 and Parent 2. In each case, equal sets of alleles are used in the crossovers to create two new children, Child 1 and Child 2, respectively.

5.5.2.2 Double crossover
The operation of double crossover is similar to single crossover except that the crossover takes place at two positions between mates.

$$x' = \langle x_1, \ldots, x_i, \quad y_{i+1}, \ldots, y_j, \quad x_{j+1}, \ldots, x_n \rangle$$
$$y' = \langle y_1, \ldots, y_i, \quad x_{i+1}, \ldots, x_j, \quad y_{j+1}, \ldots, y_n \rangle \tag{5.7}$$

The indices $i, j \in N_{n-1}; i < j$

5.5.2.3 Mutation
The two steps, 'selection' and crossover, produce offspring or a new population filled with individuals consisting of crossover individuals and others copied into the new population. Due to this step, the new population could contain exact replicas of the old population from the parents. In nature, mutation could take place and it is normal too; in genetic algorithms, mutation is

allowed to ensure that the individuals in the new population are not exactly the same. Mutation is necessary to ensure that there is diversity in the population. Mutation is the process whereby a chromosome undergoes a change in which one (or more genes) is replaced with new gene. This is illustrated in the next equation for a single-position mutation. The chromosome in the given example has one of its genes replaced with q. In the equation, the gene at position i is replaced with a gene q.

$$x' = \langle x_1, \ldots, x_{i-1}, \quad q, x_{i+1}, \ldots, x_n \rangle \qquad (5.8)$$

The gene q is randomly chosen from the gene pool Q.

5.5.2.4 Inversion

Apart from mutation, inversion could take place within the population. Given the chromosome $x = \langle x_1, x_2, \ldots, x_n \rangle$ and the integers $i, j \in N_{n-1}; i < j$ as the inversion positions, then the chromosome is inverted to

$$x' = \langle x_1, \ldots, x_i, \quad x_j, x_{j-1}, \quad x_{i+1}, \quad x_{j+1}, \ldots, x_n \rangle \qquad (5.9)$$

In most genetic algorithms, mutation and inversion operations are not involved. Their roles are however to produce new chromosomes. The new chromosomes are created to avoid the process descending to local minima without using fitness functions. The genes involved in mutation and inversions are chosen using very small probabilities.

Therefore, algorithmically, the basic structure of a genetic algorithm is presented in the following flowchart (Figure 5.6).

Each iteration in the figure produces a new generation. New generations are affected by crossover and mutation. Each generation has at least a highly fit member of the generation. There remains the problem of which criteria to use in terminating the iterations. In the rest of the chapter, the concepts developed in the previous sections are applied to different problems. Each problem reveals a method of solution and choice of fitness function and crossover.

5.6 Maximizing a Function of a Single Variable

This example originates from Goldberg [1] and is adapted by Carr [2]. The problem is to maximize the quadratic function:

$$f(x) = 3x - \frac{x^2}{10} \qquad (5.10)$$

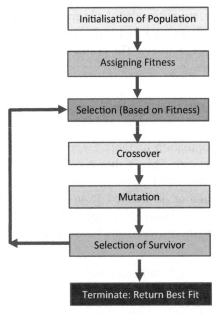

Figure 5.6 Genetic algorithm steps.

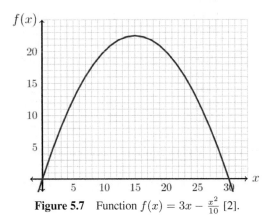

Figure 5.7 Function $f(x) = 3x - \frac{x^2}{10}$ [2].

The variable x lies between 0 and 31 as shown in Figure 5.7. To encode all values of the variable, five-bit chromosomes are required. This is because there are 32 values represented as binary numbers. The range of numbers is therefore from 00000 to 11111. Since x has a limit 10 placed on it, it is reasonable to pick 10 numbers randomly out of the 32 as the initial population.

From this table, we notice that the average score is 17.37. The cumulative sum of the fitness function x is 173.7. The minimum and maximum values are 8.1 and 22.4, respectively. The column on probability is computed with the expression:

$$p(x_i) = \frac{f(x_i)}{\sum_{k=1}^{10} f(x_k)}; \quad 1 \le i \le 10 \tag{5.11}$$

From the initial population in Table 5.1, pairs of chromosomes are selected for mating or crossover. This is shown in Table 5.2. With 10 chromosomes and pairs to mate, five mating groups are shown.

The new population is derived from the parent population by mating pairs of chromosomes. This is followed with mutation of two offspring. The colours demarcate where crossovers have taken place and the red bits where

Table 5.1 Initial population

Chromosome Index (i)	Initial Population	Value of x	Fitness $f(x)$	Probability of Selection $p(x_i)$
1	10101	21	18.9	0.10881
2	00111	7	16.1	0.09269
3	11001	25	12.5	0.07196
4	10001	17	22.1	0.12723
5	10110	22	17.6	0.10132
6	11000	24	14.4	0.08290
7	10100	20	20	0.11514
8	10000	16	22.4	0.12896
9	10010	18	21.6	0.12435
10	11011	27	8.1	0.04663

Table 5.2 Mating, crossover, mutation and new population

Chromosome Index (i)	Mating Pairs	New Population	Value of x	Fitness $f(x)$	Probability of Selection $p(x_i)$
1	10101	10001	17	22.1	0.12458
3	11001	11101	25	12.5	0.07046
2	00111	10110	21	18.9	0.10654
5	10110	10111	23	16.1	0.09076
4	10001	10100	20	20	0.11274
8	10000	10001	17	22.1	0.12458
6	11000	10100	20	20	0.11274
7	10100	11000	24	14.4	0.08117
9	10010	10011	19	20.9	0.11781
10	11011	11010	26	10.4	0.05862

mutations have occurred. Using a fairly high mutation probability of 0.025 from 50 bits ($50 \times 0.025 = 1.25$ bits), two bits are expected to mutate.

The maximum fitness of the new population is 177.4. The average fitness for the population is 17.74, which is higher than that for the parent population. Two chromosomes have mutations, and the positions of the mutations are marked with red bits. The bits are flipped in value from zero to one. Compared to their parents, this population has higher fitness sum and bigger average fitness. The above algorithm is repeated until there is a stopping point. The stopping point could be that there is no more significant changes in fitness of new generations and then stopped.

5.7 Continuous Genetic Algorithms

Many applications in data analytics with sensors involve floating point numbers. All the previous examples given so far relate to integers and indeed binary numbers. In this section, we provide examples on how to use genetic algorithms when the data is floating point. For such cases, the chromosome is an array of floating point numbers instead of integers. Therefore, the precision of the solution is only a function of the computing device not the algorithm. The dimensionality of the problem defines the size of the array. If the dimension is N_r, then the chromosome C_r also have N_r variables of the form:

$$C_r = \begin{bmatrix} p_1, & p_2, \ldots, p_{N_r} \end{bmatrix} \tag{5.12}$$

Each instance of the ith variable is p_i in Equation (5.12) and is a real number instead of an integer or bit. In the next sections, we provide two examples in the use of continuous genetic algorithms following the examples in [2]. The first example is to find the point of lowest elevation in a topographical map. The second example is to find the temperature distribution in an agricultural field as given by temperature and location sensors.

5.7.1 Lowest Elevation on Topographical Maps

The system parameters include the longitude (x) in the range 0 to 10 and latitude (y) in the range 0 to 10. The topography of the area is represented with a sinusoidal function given by the expression:

$$f(x, y) = x \sin(4x) + a y \sin(2y) \tag{5.13}$$

The constant 'a' takes values for different elevations. For this example, $a = 1.1$. The function $f(x, y)$ is clearly the fitness function. The system chromosome needs to be defined. Since 'a' is a constant, it should not feature as part of the chromosome. The chromosome contains two variables x and y defined by the longitude and latitude. Therefore, the chromosome is

$$C_r = [x, y]; \quad 0 \le x \le 10; \quad 0 \le y \le 10 \tag{5.14}$$

The choice of the length of chromosome is important as it determines how fast the system runs. Since the variables are real numbers, there is no clear size of the chromosome. One possible choice is to use all the discrete values of the longitude and latitude, which means the size of the chromosome could be up to 22. This is too large a size for the chromosome. In [2], the size is limited to 12 and we use the same here. The mutation rate is 0.2, and the number of iterations is 100. The speed of convergence of the algorithm depends on these values. Tables 5.3 and 5.4 show two initial populations. The first table has a population of 12 and the second 6. The second table is a selection from Table 5.3.

In Table 5.4, selections are made from Table 5.3 to provide a surviving 50% population.

Note that the fitness values for the populations have negative numbers. This therefore precludes the use of probabilities. Probabilities depend on summations of fitness to define them. How should the fitness function be defined? Table 5.4 ranks the individuals in terms of the values of their fitness from the most negative to the most positive. This ranking leads to the following probability function for the nth chromosome. The variable N_{keep}

Table 5.3   Initial population [2]

x	y	Fitness $f(x, y)$
6.7874	6.9483	13.5468
7.5774	3.1710	−6.5696
7.4313	9.5022	−5.7656
3.9223	0.3445	0.3149
6.5548	4.3874	8.7209
1.7119	3.8156	5.0089
7.0605	7.6552	3.4901
0.3183	7.9520	−1.3994
2.7692	1.8687	−3.9137
0.4617	4.8976	−1.5065
0.9713	4.4559	1.7482
8.2346	6.4631	10.7287

Table 5.4 Surviving population with 50% selection [2]

Rank	x	y	Fitness $f(x, y)$
1	7.5774	3.1710	−6.5696
2	7.4313	9.5022	−5.7656
3	2.7692	1.8687	−3.9137
4	0.4617	4.8976	−1.5065
5	0.3183	7.9520	−1.3994
6	3.9223	0.3445	0.3149

is the total number of organisms kept in the chromosome.

$$P_n = \frac{N_{keep} - n + 1}{\sum_{i=1}^{N_{keep}} i} = \frac{6 - n + 1}{1 + 2 + 3 + 4 + 5 + 6} = \frac{7 - n}{21} \qquad (5.15)$$

By keeping only 6 organisms, we have limited the mating groups to 3 to produce the next population.

It was easy to implement crossover from bit strings. Unfortunately, in this example, we do not have bit strings anymore. How should we implement crossover in continuous genetic algorithm? The approach adopted was initially given by Haupt [3]. Consider two parent chromosomes the 'hill' and 'valley'. For these two sets of parents let

$$h = [x_h, y_h] \quad \text{and} \quad v = [x_v, y_v] \qquad (5.16)$$

Define a variable β such that $0 \le \beta \le 1$ and create new offspring by using a ratio for mating as

$$\left. \begin{array}{l} x_{new1} = (1 - \beta)x_h + \beta x_v \\ x_{new2} = (1 - \beta)x_v + \beta x_h \end{array} \right\} = \begin{bmatrix} 1 - \beta & \beta \\ \beta & 1 - \beta \end{bmatrix} \begin{bmatrix} x_h \\ x_v \end{bmatrix} \qquad (5.17)$$

The matrix form, which we introduce here, is easier to remember as in equation (5.17). The second parameter is inherited from their parents without modifications, and the offspring are:

$$\left. \begin{array}{l} offspring_1 = [x_{new1}, y_h] \\ offspring_2 = [x_{new2}, y_v] \end{array} \right\} \qquad (5.18)$$

Example 3

Consider chromosome1 $= [7.5774, 3.1710]$ and chromosome2 $= [7.0605, 7.6552]$ and define the parameter $\beta = 0.375$, find the new offspring?

From Equation (5.17):

$$\begin{bmatrix} x_{new1} \\ x_{new2\ v} \end{bmatrix} = \begin{bmatrix} 1 - \beta & \beta \\ \beta & 1 - \beta \end{bmatrix} \begin{bmatrix} x_h \\ x_v \end{bmatrix} = \begin{bmatrix} 0.25 & 0.375 \\ 0.375 & 0.25 \end{bmatrix} \begin{bmatrix} 7.5774 \\ 7.0605 \end{bmatrix} = \begin{bmatrix} 4.5420 \\ 4.6067 \end{bmatrix}$$

The offspring are:

$$\begin{bmatrix} \textit{offspring}_1 \\ \textit{offspring}_2 \end{bmatrix} = \begin{bmatrix} 4.5420, & 3.1710 \\ 4.6067, & 7.6552 \end{bmatrix}$$

The mutation procedure when a chromosome mutates replaces its value with a random number chosen between 0 and 10. In general, mutation is undertaken also using many different methods. We have shown this example for one iteration. Normally, repeat the procedure many times to find the new generations. Observe that continuous genetic algorithm does not need the decoding step and hence it is a lot faster to implement.

5.7.2 Application of GA to Temperature Recording with Sensors

Recording of temperatures, humidity, location and environmental conditions is a very popular requirement in precision agriculture. Other conditions soil conditions recorded with sensors include soil water content, nitrogen, potassium and sodium chloride. In this section, we demonstrate how to use genetic algorithm for predicting temperature distribution in a farm. The temperature distribution is given by the expression

$$T(x, y) = T_0 + (T_1 - T_0) \exp \left(\frac{(x - x_0)^2 + (y - y_0)^2}{2\lambda^2} \right) \qquad (5.19)$$

Define the system parameter

$$g = [T_0, \ T_1, \ x_0, \ y_0, \ \lambda] \qquad (5.20)$$

This is used to create the set of chromosomes and the initial population.
We need to minimize the function

$$\Phi^2(g) = \sum_{i=1}^{N} \left(\frac{T_i - T(x_i, y_i | g)}{30} \right)^2 \qquad (5.21)$$

The encoding to allow GA to be used needs conversion of the system vector of parameters g to chromosomes. In other words, g is an organism (temperature and position set). By converting the vector to binary sets, we obtain the system chromosomes or DNA. Five values therefore need to be encoded for each value of temperature at each location. The fitness function is related to the summation expression above. The objective is to minimize the error between measured values at a location referenced to a temperature value T_i. The fitness function therefore is

$$f(g) = \frac{1}{\Phi^2(g)} = \frac{1}{\sum_{i=1}^{N} \left(\frac{T_i - T(x_i, y_i|g)}{30}\right)^2} \tag{5.22}$$

We would therefore need to minimize the temperature error at a location and in doing so maximize $f(g)$.

In general, the genetic algorithm becomes useful when the search space is large with a large number of parameters. This situation occurs in Big Data analysis. Genetic algorithm is applicable over integer and floating-point values, which is a huge benefit. Generally, good solutions are found using genetic algorithms, and it is faster and efficient as well. The mathematics is simple and applies to real-world problems. Thus, the field of application is diverse.

References

[1] David E. Goldberg, Genetic Algorithms in Search Optimization and Machine Learning, Addison-Wesley, 1989.
[2] Jenna Carr, "An Introduction to Genetic Algorithm", 2014.
[3] R.L. Haupt and S.E. Haupt, Practical Genetic Algorithm, 2nd Edition, Hoboken: Wiley, 2004.

6

Calculus on Computational Graphs

6.1 Introduction

Fast, accurate and reliable computational schemes are essential when implementing complex systems required in deep learning applications. One of the techniques for achieving this is the so-called computational graph. Computational graphs divide down a complex computation into small and executable steps which could be performed quickly with pencil and paper and better still with computers. In most cases, loops that require repeating of the same algorithm but wastes time computationally due to processing of loop times become a lot easier to handle. Computational graphs ease the training of neural networks with gradient descent algorithm making them many times faster than traditional implementation of neural networks.

Computational graphs have also found applications in weather forecasting by reducing the associated computation time. Its strength is fast computation of derivatives. It is known also by a different name of "reverse-mode differentiation."

Beyond its use in deep learning, backpropagation is a powerful computational tool in many other areas like weather forecasting and analysis of numerical stability. In many ways, computational graph theory is similar with logic gate operations in digital circuits where dedicated logic operations are undertaken with logic gates such as AND, OR, NOR and NAND operations in the implementation of many binary operations. While the use of logic gates lead to complex systems such as multiplexers, adders, multipliers and more complex digital circuits, computational graphs have found their way into deep learning operations involving derivatives of real numbers, additions, scaling and multiplications of real numbers by simplifying the operations.

6.1.1 Elements of Computational Graphs

Computational graphs are useful means of breaking down complex mathematical computations and operations into micro computations and thereby make it a lot easier to solve them in a sequential manner. They also make it easier to track computations and to understand where solutions break down. A computational graph is a connection of links and nodes at which operations take place. Nodes represent variables and links are functions and operations.

The range of operations include addition, multiplication, subtraction, exponentiation and a lot more operations herein not mentioned. Consider Figure 6.1, the three nodes represent three variables a, b and c. The variable c is the result of the operation of the function f on a and b. This means we can write the result as

$$c = f(a, b) \tag{6.1}$$

Computational graphs allow nesting of operations which allow solving more complex problems. Consider the following nesting of operations with this computational graph in Equation (6.2).

Clearly, from these three operations, we see that it is a lot easier to undertake the more complex operation when y is computed as in equation

$$y = h(g(f(x))) \tag{6.2}$$

From Figure 6.2, it is apparent that the first operation is performed first $(f(x))$. This is followed by the second operation, which is $g(.)$, and lastly $h(.)$ operation.

However, from the point of view of Equation (6.2), the innermost operation in Equation (6.2) involving $f(x)$ is performed first. This is followed by the second operation, which is $g(.)$, and lastly $h(.)$ as the last operation.

Figure 6.1   Building components of a typical unit cell.

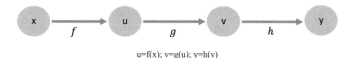

u=f(x); v=g(u); y=h(v)

Figure 6.2   Operation flow in computational graphs.

Consider the case where f is the operation of addition $(+)$. The result c is then given by the expression

$$c = (a + b)$$

when the operation is multiplication (x), the result is also given by the expression.

$$c = (a \times b)$$

Since f is given as a general operator, we may use any operator in the diagram and write an equivalent expression for c.

6.2 Compound Expressions

For computer to use compound expressions with computational graph efficiently, it is essential to factor expressions into unit cells. For example, the expression $r = p \times q = (x + y)(y + 1)$ can be reduced first to two unit cells or two terms followed by computation of the product of the terms. The product term is $r = p \times q$.

$$r = p \times q = (x + y)(y + 1) = xy + y^2 + x + y$$
$$p = x + y$$
$$q = y + 1$$

Each of the computational operation or component is created and then a graph is built out of them by appropriately connecting them with arrows. The arrows originate from the terms used to build the unit term where the arrow ends as in Figure 6.3.

This form of abstraction is extremely useful in building neural networks and deep learning frameworks. In fact, they are also useful in programming of expressions for example in operations involving parallel support vector machines (PSVM). In Figure 6.4, once the values at the root of the computational graphs are known, the solution of the expression becomes a lot easy and trivial. Take the case where $x = 3$ and $y = 4$, the resulting solutions are shown in Figure 6.5.

The evaluation of the compound expression is 35.

Figure 6.3   Building components of a unit cell.

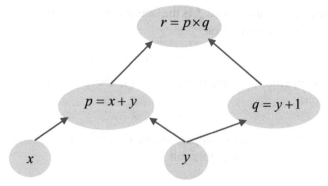

Figure 6.4   Computational graphs of compound expressions.

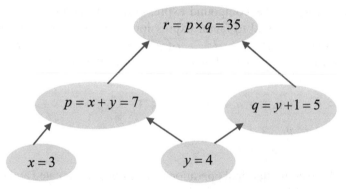

Figure 6.5   Evaluating compound expressions.

6.3 Computing Partial Derivatives

One of the areas where computational graphs are used widely in neural network applications is for computing derivatives of variables and functions in simple forms. For derivatives, it simplifies the use of the chain rule. Consider computing the partial derivative of y with respect to b where

$$y = (a + b) \times (b - 3) = c \times d$$
$$c = (a + b); \quad d = (b - 3)$$

(6.3)

The partial derivative of y with respect to b is

$$\frac{\partial y}{\partial b} = \frac{\partial y}{\partial c} \times \frac{\partial c}{\partial b} + \frac{\partial y}{\partial d} \times \frac{\partial d}{\partial b}$$

(6.4)

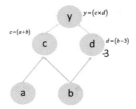

Figure 6.6 Computational graph for Equation (6.3).

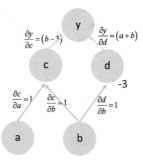

Figure 6.7 Computation of partial derivatives.

The partial computational graph covering Equation (6.3) is given in Figure 6.6. The partial derivative is given in Equation (6.4). From Equation (6.3), the partial derivatives are given as

$$
\begin{aligned}
\frac{\partial y}{\partial c} &= d = (b - 3); \\
\frac{\partial y}{\partial d} &= c = (a + b); \\
\frac{\partial c}{\partial a} &= 1; \quad \frac{\partial c}{\partial b} = 1; \quad \frac{\partial d}{\partial b} = 1; \\
\frac{\partial y}{\partial b} &= (b - 3) \times 1 + (a + b) \times 1
\end{aligned}
\tag{6.5}
$$

These derivatives are superimposed on the computational graph in Figure 6.7.

6.3.1 Partial Derivatives: Two Cases of the Chain Rule

The chain rule may be applied in various circumstances. Two cases are of interest: the linear case (Figure 6.8) and the loop case (Figure 6.9). These two cases are illustrated in this section.

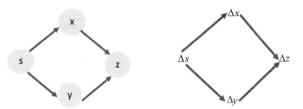

Figure 6.8   Linear chain rule.

Figure 6.9   Loop chain rule.

6.3.1.1 Linear chain rule

The objective in the linear case is to find the derivative of the output z with respect to the input x or $\frac{dz}{dx}$. The output depends recursively on both y and x and hence it is expected that the partial derivative of z will also depend on these two variables. The partial derivative of y with respect to x therefore needs to address this. From the diagram, we can write generally

$$z = f(x); \quad z = h(y); \quad y = g(x) \tag{6.6}$$

Therefore, the derivative of z with respect to x needs to be done with respect to y also and is a product of two terms:

$$\frac{dz}{dx} = \frac{\partial z}{\partial y} \times \frac{\partial y}{\partial x} \tag{6.7}$$

Thus, the derivative of z with respect to x is computed as a product of two partial derivatives of variables leading to it. This is shown in Figure 6.8.

6.3.1.2 Loop chain rule

The loop chain rule in Figure 6.9 is an application of the linear chain rule. Each loop is treated with a linear chain rule. Consider the following loop diagrams (Figure 6.9). The objective is to find the derivative of the output z using the linear chain rule along the two arms of the loop and sum them.

z is a function of s as $z = f(s)$ through two branches involving x and y, respectively. In the upper branch, $x = g(s)$ and $z = h(x)$. In the lower

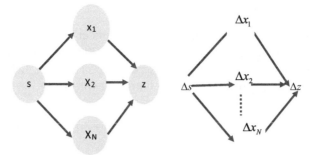

Figure 6.10 Multiple loop chain rule.

branch, $y = g(s), z = h(y)$. Two branches contribute to the value of z so that $z = p(x, y)$. Therefore, there will also be a sum of partial derivatives coming from the two branches. The derivative of x with respect to s is obtained as

$$\frac{dz}{ds} = \frac{\partial z}{\partial x} \times \frac{dx}{ds} + \frac{\partial z}{\partial y} \times \frac{dy}{ds} \tag{6.8}$$

6.3.1.3 Multiple loop chain rule

Generally, if z is computed from N loops so that it is a function of N variables like $z = k(x_1, x_2, \ldots, x_N)$, then N branches contribute to the output z. Therefore, the total derivative of z to the input is a chain of N partial derivatives.

The general partial derivative expression shown as Figure 6.10 is

$$\frac{dz}{ds} = \frac{\partial z}{\partial x_1} \times \frac{dx_1}{ds} + \frac{\partial z}{\partial x_2} \times \frac{dx_2}{ds} + \cdots + \frac{\partial z}{\partial x_N} \times \frac{dx_N}{ds}$$

$$= \sum_{n=1}^{N} \frac{\partial z}{\partial x_n} \times \frac{dx_n}{ds} \tag{6.9}$$

The general case is more suited to deep learning situations where there are many stages in the neural network and many branches are also involved.

6.4 Computing of Integrals

In the next section, we introduce the use of computational graphs for computing integrals using some of the well-known traditional approaches including the trapezoidal and Simpson rules.

6.4.1 Trapezoidal Rule

Integration traditionally is finding the area under a curve. This area has two dimensions: the discrete step in sampling of the function multiplied by the amplitude of the function at that discrete step. Thus, the function $f(x)$ may be computed using trapezoidal rule with the expression

$$\int_a^b f(x)dx \approx \frac{\Delta x}{2} [f(x_0) + 2f(x_1) + 2f(x_2) \cdots + 2f(x_{N-1}) + f(x_N)]$$

$$\approx \frac{\Delta x}{2} [f(x_0) + f(x_N)] + \Delta x \sum_{i=1}^{N-1} f(x_i) \qquad (6.10)$$

so that $\Delta x = \frac{b-a}{N}$; $x_i = a + i\Delta x$. N is the number of discrete sampling steps. Therefore, the computational graph for this integration method can be easily drawn by first computing the values of the function at N + 1 locations starting from zero to N in Δx discrete steps. Just how big N is will be determined by some error bound, which has been derived to be traditionally

$$|E| \leq \frac{K_2(b-a)^3}{12N^2}.$$

K_2 is the value of the second derivative of the function $f(x)$. This sets the bound on the error in the integration value obtained by using the trapezoidal rule for the function. Thus, once the choice of N is made, an error bound has been set for the result of the integration. This error may be reduced by changing the value of N, the number of terms in the summation. Figure 6.11 shows how to compute an integral using the Trapezoidal rule.

6.4.2 Simpson Rule

The Simpson rule for computing the integral of a function follows the same type of method as used for the trapezoidal rule with two exceptions. The summation expression is different. The number of terms N is even. The rule is

$$\int_a^b f(x)dx \approx \frac{\Delta x}{3} [f(x_0) + 4f(x_1) + 2f(x_2) + 2f(x_3) + \cdots + 2f(x_{N-2})$$

$$+ 4f(x_{N-1}) + f(x_N)]$$

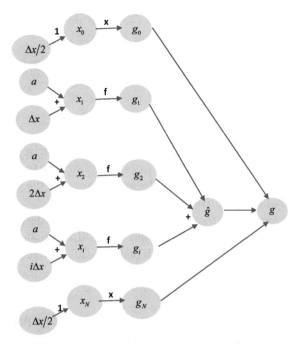

Figure 6.11   Integration using trapezoidal rule.

$$\approx \frac{\Delta x}{3} \left[f(x_0) + f(x_N) \right] + \frac{4\Delta x}{3} \sum_{i=1}^{(N/2)-1} f(x_{2i+1})$$

$$+ \frac{2\Delta x}{3} \sum_{i=1}^{(N/2)-1} f(x_{2i}) \tag{6.11}$$

so that $\Delta x = \frac{b-a}{N}$; $x_i = a + i\Delta x$. Therefore, the computational graph for this integration method can be easily drawn by first computing the values of the function at $N+1$ locations starting from zero to N in Δx discrete steps. Just how big N is will be determined by some error bound which has been derived to be traditionally

$$|E| \leq \frac{K_4 (b-a)^5}{180 N^4}$$

K_4 is the value of the fourth derivative of the function $f(x)$. This sets the bound on the error in the integration value obtained by using the Simpson rule for the function. Once the choice of N is made, an error bound has been set for the result of the integration. This error may be reduced by changing the value of N, the number of terms in the summation.

Exercise: Draw the computational graph for the Simpson Rule for integrating a function.

6.5 Multipath Compound Derivatives

In multipath differentiation as used in neural network applications, there is tandem differentiation. Results from previous steps affect the derivative of the current node. Take for example in Figure 6.12, the derivative of Y is influenced by the derivative of X in the forward path.

In the reverse path, the derivative of Z affects the derivative of Y. Let us look at these two cases involving multipath differentiation. In Figure 6.12, the weights or factors by which each path derivative affects the next node are shown with arrows and variables.

Multipath Forward Differentiation

In the discussion, we limit the number of paths to three, but with the understanding that the number of paths is limitless and depends on the application. Observe the dependence of the partial integrals on the weights from the integrals from the previous node.

In Figure 6.13, we have the derivative of Z with respect to X depends on the derivative of Y with respect to X. Notice the starting point has the derivative of a variable X to X.

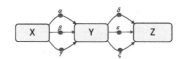

Figure 6.12   Multipath differentiation.

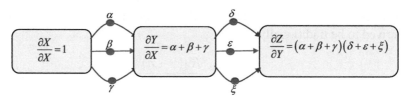

Figure 6.13   Multipath forward differentiation.

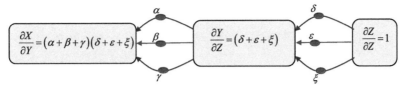

Figure 6.14 Multipath reverse differentiation.

Multipath Reverse Differentiation

In the reverse (backward) path dependence, Figure 6.14 shows that the derivative of Z with respected to X starts with the derivative of Z with respect to Z, which of course is one.

In Figure 6.14, there is a tandem of three paths from stage 1 and another three paths to the end node, making a total of 9 paths. These paths for the forward and reverse partial differentiations are given by the product $(\alpha + \beta + \gamma)(\delta + \varepsilon + \xi)$.

7

Support Vector Machines

7.1 Introduction

Support vector machines (SVM) are classifier algorithms. SVM combines the best features of linear classification methods with optimization techniques to identify which data belongs to one class or the other. This form of supervised learning algorithm was introduced by Vladimir Vapnik in 1992. Since introduction, it has become more and more popular and ranks in line with other learning algorithms like neural networks. Comparatively, it is a lot easier to train an SVM than neural networks. It also does not possess local minima unlike in NN where in gradient descent local minima could result and impact on the convergence performance of NN. For these reasons and others, it has found application in hand-writing digit recognition. Although NN may be used for digit recognition as well, it requires elaborate algorithmic overhead. Other areas where SVM has been widely applied include data mining, hand-writing recognition, bioinformatics, proteomics, medicine, image classification and biosequence analysis.

Classification of data using SVM requires two stages of operation. The first stage is learning. During this stage, labelled data is analyzed to learn a mapping from x to y, where x is the set of data and y the class set. The aim of this stage is to build a classifier. The second stage is the prediction stage using the classifier obtained from the first stage to predict which class the inputs belong to. To determine the classifier or model, which is a convex optimization problem, a local minimum is sought. The spread in data in an R^2 space leads to two popular methods of analysis, the linearly separable case and the nonlinear case. Of these, the linearly separable case is the easiest and usually SVM analysis using this method is rare. For the linearly separable case, classification of data pairs is almost perfect with little or no errors. Assume that the training set are pairs of two-dimensional data labelled as

$\{(x_1, y_1), (x_2, y_2), \cdots, (x_n, y_n)\}$. The pairs of data belong to the two classes $y \in \{+1, -1\}$. During the training stage, the classifier learns the parameters of the system, which are w and b. They are used to compose a decision function $f(x) = sign(w^T \cdot x + b)$, where w is a set of weights and b is the bias. Once the SVM has learned the weights and b from the training points, they are ready to be used to produce outputs corresponding to unknown inputs. These outputs indicate which classes the input data belongs to. This is based on the decision function. In the linearly separable data, the classification solves the quadratic optimization problem to minimize $\frac{1}{2}\|w\|^2$ subject to the constraint $y_i(w^T x_i + b) \geq +1; \quad i = 1, 2, \cdots, N$. This is basically saying that we want to find the best or optimal weights for classifying all the data in the data sets. When we have a parallel SVM, each parallel set needs to find its own optimal weights as well. This chapter is only on the traditional single SVM. Its expressions will be derived in this chapter.

In the tutorial-type discussions in this chapter, we cover the basis of SVM and linear separable types of support vector algorithms and nonlinear types. Kernel algorithms are included with various kernel functions discussed. Some applications of support vector machines are discussed towards the end of the chapter.

This chapter also uses vector theory heavily. Therefore, it is advised that a reader not familiar with vectors take time to first study vectors and then to apply vector calculus. Specifically, vector concepts including length of vectors, vector projections, scalars, dot products, cross products and norms should be reviewed.

7.2 Essential Mathematics of SVM

To illustrate what follows in the coming sections, consider as an example, application of SVM in the health sector. A patient in a clinic is given attributes that define the patient. The attributes are related to the type of symptoms to an illness the patient has. Consider the patient has high temperature and blood pressure. These are measurable variables which are components of a vector $\vec{x}$, where the vector describing the combination of temperature and blood pressure is $x = (T, p)$. We wish to classify patients in terms of the type of illness. In its simplest form, we use linear regression with a set of weights and an offset b given by the output function $y = \vec{w} \cdot \vec{x} + b$, where $\vec{w}$ are regression weights.

We will consider vector spaces including Euclidean ($n = 2$) and therefore represent real numbers in such spaces as R^n, where $n = 2$ for a two-dimensional (2D) space and $n = 3$ for a three-dimensional (3D) space.

Support vector machines use planes a lot to classify data into clusters. It is therefore essential that we derive functional understanding of the relationships between planes and vectors, and how they interact with each other. In this section the mathematics behind the relationships between vectors and planes are presented and explained. These techniques are derived from the works of Statnikov et al. [1] on support vector machines.

7.2.1 Introduction to Hyperplanes

In two dimensions (the x-y plane), the traditional equation for a line is well known to be the expression $y = mx + c$, where y is the y-coordinate, x is the x-coordinate for the line and c is the intercept on the y-axis when $x = 0$. The same equation for a line is often written in a different form as $ax + by + c = 0$. It is easy to show that the two expressions lead to the same solution when we cast the general equation as

$$y = -\frac{c}{b} - \frac{a}{b}x \tag{7.1a}$$

Setting $c_0 = -\frac{c}{b}; m = -\frac{a}{b}$ leads to the same expression for a line in a two dimensional space. Suppose it were possible to extend this concept into more dimensional spaces. Let us first start with a three dimensional (3D) space. With foresight into what we will be discussing in the next sections, we will write the expression for a line in two dimensions first by introducing w as a general variable to be carried over to higher dimensional spaces. The line in 2D therefore is $w_1 x_1 + w_2 x_2 + w_0 = 0$. Doing so changes nothing in terms of the expression for a line. The only difference is that we now represent the axes in a two-dimensional space with x_1 and x_2. Let us extend this concept to 3D. The traditional expression for a line in 3D is $ax + by + cz + d = 0$ and when replaced with w, the expression for a line becomes $w_0 + w_1 x_1 + w_2 x_2 + w_3 x_3 = 0$. In 3D, the 'line' is actually a plane which divides the space into two sub-spaces below and above the plane. We call this plane a hyperplane just to keep in view of the need to have a general concept which can be extended into higher dimensional spaces. Therefore in n-dimensional spaces the expression for a hyperplane is $w_0 + w_1 x_1 + w_2 x_2 + w_3 x_3 + \cdots + w_n x_n = 0$. Note that no mention has been made on the possibility for viewing the hyperplanes in these n-dimensional space as that is not the objective here; rather the objective is the ability to recognize the concept for representing hyperplanes mathematically. In general we could have written the expression for the hyperplane in n-dimensions as

$$w_0 + \sum_{i=1}^{n} w_i x_i = 0 \qquad (7.1b)$$

This expression is actually written for the sake of its use in SVM as a vector. Traditionally vectors are written as columns of data as $\begin{bmatrix} w_1 \\ w_2 \\ \vdots \\ w_n \end{bmatrix}$. This does not allow direct multiplication with another column vector $\begin{bmatrix} x_1 \\ x_2 \\ \vdots \\ x_n \end{bmatrix}$ because it is impossible to multiply two column vectors of size $(n \times 1) \times (n \times 1)$. Therefore, the expression for hyperplanes is normally written in a form which allows multiplication using the transpose operation. This leads to the expression $\vec{w}^T \cdot \vec{x} + b = 0$. Writing the expression in this form means we can multiply a row vector and a column vector of sizes $(1 \times n)(n \times 1)$ to obtain a scalar result.

In SVM theory, a hyperplane in 2D is a defined linear decision surface that divides a space into two parts in an R^2 space. The equation for the hyperplane shown in Figure 7.1 is $\vec{w}^T \cdot \vec{x} + b = 0$. This equation is derived in this section.

In Figure 7.1, equations of hyperplanes are determined by three vectors in the R^3 space with origin at (0,0,0). The vector $\vec{w}$ passes through the plane at point P_0. This vector or normal to the hyperplane has three components in the x, y and z directions such that $\vec{w} = (\vec{i}w_x, \vec{j}w_y, \vec{k}w_z)$ and $(\vec{i}, \vec{j}, \vec{k})$ are unit vectors. They also can be used to define the plane equation. From Figure 7.1, the following vectors can be identified:

$$\vec{x} = O\vec{P}, \quad \vec{x}_0 = O\vec{P}_0 \qquad (7.2a)$$

Consider the orthogonal vector to $\vec{w}$ in Equation (7.2b)

$$\vec{x} - \vec{x}_0 = P\vec{P}_0 \qquad (7.2b)$$

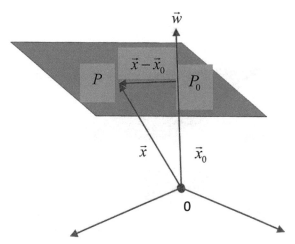

Figure 7.1 Definition of hyperplane.

Therefore, using the principles of orthogonality, we can write the dot product using Equations (7.2a) and (2) as being zero. This is written in terms of the row vector $\vec{w}^T$:

$$\vec{w}^T.(\vec{x} - \vec{x}_0) = 0 \qquad (7.3)$$

Hence, $\vec{w}^T \cdot (\vec{x} - \vec{x}_0) = \vec{w}^T \cdot \vec{x} - \vec{w}^T \cdot \vec{x}_0$. Let $b = -\vec{w}^T \cdot \vec{x}_0$. Therefore, this hyperplane equation simplifies to

$$\vec{w}^T \cdot (\vec{x} - \vec{x}_0) = \vec{w}^T \cdot \vec{x} + b = 0 \qquad (7.4)$$

The distance of the plane from the coordinate origin is defined by the expression: $d = \frac{b}{\|w\|}$. Let us solve an example to illustrate this.

Example 1: Consider a hyperplane in an R^3 space. The point P lying on the plane is defined by the triplet $P(3, -1, 4)$ and the normal vector to the plane is $(3, -2, 5)$. (i) Find the distance of the plane from the origin (ii) write the equation of the hyperplane.

Solution

(i) The distance of the plane from the origin is given by the expression is the dot product of w and x:

$$k = \vec{w}^T \cdot \vec{x} = (3, -1, 4) \cdot (3, -2, 5)$$
$$= (3 \times 3 + (-1) \times (-2) + 4 \times 5) = 9 + 2 + 20 = 31$$

(ii) The equation of the hyperplane is thus, $\vec{w}^T \cdot \vec{x} + b = 0$, where $b = -31$.

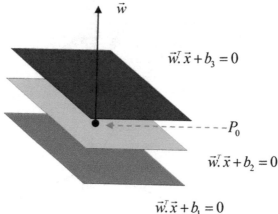

Figure 7.2   Parallel hyperplanes.

7.2.2 Parallel Hyperplanes

Figure 7.2 shows three parallel hyperplanes defined in a three-dimensional space. They are separated from each other by distances b defined in the plane.

Figure 7.2 shows three parallel hyperplanes at distances b_1, b_2 and b_3 from the origin (0,0,0). In the next section, we show how to determine the distances between two hyperplanes.

7.2.3 Distance between Two Parallel Planes

Consider the distance between two parallel planes as shown in Figure 7.3. The three vectors in Figure 7.3 are related by the expression

$$\vec{x}_2 = \vec{x}_1 + c\vec{w} \tag{7.5}$$

The distance between the two planes is

$$d = c\vec{w} = |c| \, \|\vec{w}\| \tag{7.6}$$

Using the equation for x_2, we can rewrite the equation for the hyperplane as

$$\vec{w}^T.\vec{x}_2 + b_2 = \vec{w}^T.(\vec{x}_1 + c\vec{w}) + b_2 = 0 \tag{7.7}$$

Therefore, inserting the expression for the distance d, this equation becomes:

$$\vec{w}^T.\vec{x}_2 + b_2 = \vec{w}^T.(\vec{x}_1 + |c| \, \|\vec{w}\|) + b_2 = 0$$
$$= (\vec{w}^T.\vec{x}_1 + b_1) + b_2 - b_1 + c \, \|\vec{w}\|^2 = 0 \tag{7.8}$$

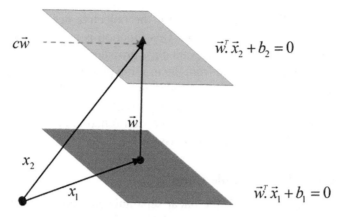

Figure 7.3 Two parallel hyperplanes.

Since $\vec{w}^T \cdot \vec{x}_1 + b_1 = 0$, the above equation reduces to Equation (7.8):

$$b_2 - b_1 + c\|\vec{w}\|^2 = 0$$
$$c = \frac{b_2 - b_1}{\|\vec{w}\|^2} \tag{7.9}$$

Therefore

$$d = |c|\,\vec{w} = \frac{|b_2 - b_1|}{\|\vec{w}\|} \tag{7.10}$$

The distance between the two planes depends on the vector w passing through the planes. In general, it is observed that the orthogonality expression $\vec{w} \cdot \vec{x}$ between a vector on the plane and the vector w is a special case of a plane located at the origin of an R^n coordinate system when we set the distance $b = 0$.

7.3 Support Vector Machines

7.3.1 Problem Definition

Consider a set of N vector points $X = \{x_1, x_2, \cdots, x_N\}$ with each component vector of length x_i. We assume that the vector points belong to two classes, the positive class "+1" and the negative class "−1" class. We also have a training vector set $X_T = \{(x_1, y_1), (x_2, y_2), \cdots, (x_N, y_N)\}$. The objective for using a training vector set is to use the set to determine the 'optimum' hyperplane $w.x + b = 0$, which separates the two classes

(Figure 7.3a). Therefore, we look for the objective decision function $f(x)$ such that $f(x) = sign(w.x + b)$. This function is used to classify any new data point received into the relevant class provided the weights w are known. The weights are determined from the training phase of SVM. In other words, for each test point x, the decision function $f(x)$ is used based on the optimum weights to return a sign. If the sign is "+1", the input data is in the positive class. If however the sign from the decision function is "−1", the input data is classified to be in the negative class.

7.3.2 Linearly Separable Case

Two broad cases of SVM, the linear separable case and the non-linear case, are considered in the discussions. In the linearly separable SVM, it is assumed that after training, the SVM is able to clearly use linear separators to classify input data into classes without errors. The decision function is

$$f(x) = sign\left(\vec{w}^T.\vec{x} + b\right) \tag{7.11}$$

In Figure 7.4, there are many hyperplanes that could be drawn without knowing which one is optimum hyperplane. Therefore, assessing how far each hyperplane is with respect to the desired optimum hyperplane is crucial in classifying the two groups of vector points efficiently.

As shown in Figure 7.5, there are infinite hyperplanes to choose from. Initially, it is not obvious which hyperplane is the optimum separating hyperplane to use. Choosing a wrong hyperplane may mean that when a new point

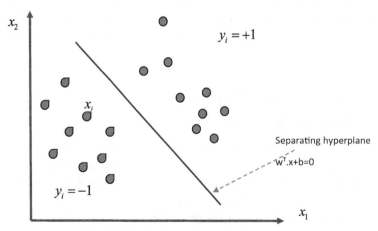

Figure 7.4   Optimum separating hyperplane.

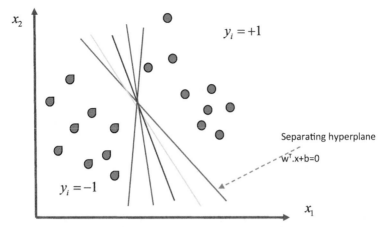

Figure 7.5 Numerous separating hyperplanes.

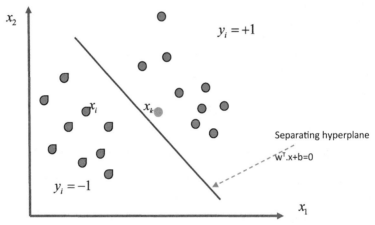

Figure 7.6 Classification of new vector point x_k correctly into the "+1" class.

is received belonging to either class, it is possible that we could misclassify it completely. When the hyperplane is far from both data classes, it becomes easier to correctly classify a new vector point such as x_k in Figure 7.6.

In general, it is desirable that the hyperplane be as far as possible from the closest support vector points in the classes. In other words, there should be no confusion which class data belongs to, or the Euclidean distance from the hyperplane to the support vector points should be maximized. When the optimal hyperplane is obtained, the distance between the hyperplane and the closest support vector points in the negative and positive classes should be equal. This calls for the canonical form of hyperplanes:

$$\left.\begin{array}{l} w^T x + b = +1 \\ w^T x + b = -1 \end{array}\right\} \tag{7.12}$$

The optimal hyperplane equation is in the middle of the two linear graphs in Equation (7.12), which is

$$w^T x + b = 0 \tag{7.13}$$

Implicitly, we have said therefore that to find the optimum hyperplane, the set of weights which determine it need to be determined using some optimization scheme, which we are yet to describe. We will do so in a later section of this tutorial chapter. We will in doing so divide our data set into a training set and the rest as the test set for classification (see Figure 7.3a).

The best hyperplanes are defined by two planes and the optimum hyperplane in their middle defining a 'margin' between them. The objective is to maximize the margin m. The margin is twice the distance from the hyperplane to the support vectors for either of the classes (m = 2d). The support vectors are defined as the vector points lying on or closet to the separating hyperplanes. In the next section, we derive the expression for the separating distance from the separating hyperplane to a training vector x_i.

7.4 Location of Optimal Hyperplane (Primal Problem)

Determining the location of the optimal hyperplane is fundamental to separating the data sets into their clusters well. In this section, this is done by first deriving the so-called margin. The margin is defined by the hyperplane and two support vector planes. The support vectors lie on the support vector hyperplanes shown in Figure 7.7.

7.4.1 Finding the Margin

The margin is defined as the width of the separating hyperplanes used to distinguish the classes in the data set (Figure 7.7). On either side of the margin are two hyperplanes. In Figure 7.7, one of them demarcates the "+1" class and the other the "−1" class. In the middle of the margin is the dividing hyperplane defined by the equation $w^T x + b = 0$.

Let x_i be a training vector at distance ρ_i from the hyperplane. This is shown in Figure 7.8. The hyperplane $w^T x + b = 0$ is orthogonal to the weight

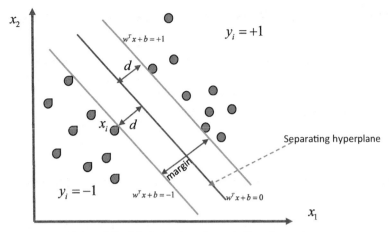

Figure 7.7 Maximization of margin from hyperplane.

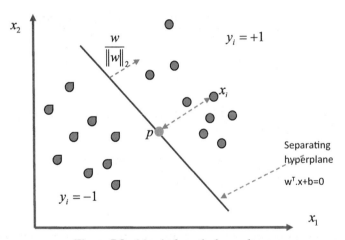

Figure 7.8 Margin from the hyperplane.

unit vector $\frac{w}{\|w\|}$, where $\|w\|$ is the Euclidean norm of the weight w. Note that $\rho_i \in R$ (a real number), $(x, w) \in R^m$ (lie in an m-dimensional space), and $b \in R$ (is also a real number). These real numbers define the parameters for the SVM. The point p on the hyperplane satisfies the equation for the hyperplane. In other words

$$w^T p + b = 0 \qquad (7.14)$$

7.4.2 Distance of a Point $\vec{x}_i$ from Separating Hyperplane

The distance $\rho_i(w, b; \vec{x}_i)$ of a point $\vec{x}_i$ is determined from the set of weights that determine the optimal hyperplane. The optimal hyperplane is determined in this section. The expression for the optimal weights is also derived.

Consider the training vector point which lies on a non-optimal hyperplane. The distance from point p to the training vector is

$$p = x_i - \rho_i \cdot \frac{w}{\|w\|_2} \tag{7.15}$$

By substituting Equation (7.15) into Equation (7.14), we obtain the following expression:

$$w^T p + b = w^T \left(x_i - \rho_i \cdot \frac{w}{\|w\|_2} \right) + b = 0 \tag{7.16}$$

This gives the solution

$$\rho_i(w, b) = \frac{(w^T x_i + b) \|w\|_2}{w^T w} = \frac{(w^T x_i + b) \|w\|_2}{\|w\|_2^2}$$
$$= \frac{(w^T x_i + b)}{\|w\|_2} \tag{7.16a}$$

By definition, $w^T w = \|w\|_2^2$. Therefore, the distance of any point from the hyperplane is determined as:

$$\rho_i = \frac{\|w^T \cdot x_i + b\|}{\|w\|_2} = \frac{\|w^T \cdot x_i + b\|}{\sqrt{\sum\limits_{i=1}^{N} w_i^2}} \tag{7.17}$$

The optimization problem therefore becomes, for all the training data sets to define the criteria for minimum hyperplane distance from the support vectors as

$$d = \min_{i \in 1, \cdots, n} \rho_i = \min_{i \in 1, \cdots, n} \left[\frac{\|w^T \cdot x_i + b\|}{\|w\|} \right] \tag{7.18}$$

This result is true for all same scaled w and b. Fundamentally, it is clear that the margin defined by the hyperplane in the middle is the product

$$m = 2d = 2 \min_{i \in 1, \cdots, n} \rho_i = 2 \min_{i \in 1, \cdots, n} \left[\frac{\|w^T \cdot x_i + b\|}{\|w\|} \right] \tag{7.19}$$

7.4.2.1 Margin for support vector points

For all support vector points x_s obeying the expression $\left|w^T x_s + b\right| = 1$. This is true for support vector points in either of the classes. Hence, from Equation (7.19), we can write that

$$m = 2d = 2 \min_{s \in 1, \cdots, S} \rho_s = 2 \min_{s \in 1, \cdots, S} \left[\frac{1}{\|w\|} \right] \tag{7.20}$$

The factor 2 comes from the separation between the 'negative' and 'positive' hyperplanes $w^T x + b = -1$ and $w^T x + b = +1$, respectively. This means that for such points

$$m = 2d = \frac{2}{\|w\|_2} = \frac{2}{\sqrt{\sum_{i=1}^{N} w_i^2}} \tag{7.21}$$

Thus, the optimum hyperplane is defined exclusively by the support vectors. In other words, the margin is defined exclusively by the norm of w. The problem has therefore been transformed into an optimization problem. In the next section, we pursue this line of thinking further. Equation (7.21) is fundamental to linear SVM.

7.4.3 Finding Optimal Hyperplane Problem

7.4.3.1 Hard margin

We have so far obtained a solution to the margin in an SVM and said in the previous section that the optimal solution or the margin is given by the expression

$$m = \frac{2}{\|w\|} \tag{7.21a}$$

In the rest of the analysis, we seek to maximize this solution for the SVM subject to the following constraints.

Maximize m

$$m = \frac{2}{\|w\|} \tag{7.22}$$

Subject to the constraints that

$$y_i = +1; \boldsymbol{w}^T \boldsymbol{x}_i + b \geq +1 \tag{7.23a}$$

$$y_i = -1; \boldsymbol{w}^T \boldsymbol{x}_i + b \leq -1 \tag{7.23b}$$

In other words, in the positive class, the Euclidean distances for the vectors from the support vector hyperplane is more than or equal to one. To minimize the objective function, it is required to

Minimize:

$$\phi(w) = \frac{\|w\|}{2} \tag{7.24}$$

This is the expression for the optimal hyperplane, which separates the data points into their classes. This solution is:

Subject to the constraint

$$y_i \left(\boldsymbol{w}^T \boldsymbol{x}_i + b \right) \geq +1 \tag{7.25}$$

It minimizes $\phi(w)$. It is also independent of b. When Equation (7.25) is satisfied, when the value of b changes, the hyperplane is also moved in the normal direction. This means that the margin remains the same. This is the hard margin solution. Notice that Equation (7.25) is true for the two cases when y_i is positive or negative 1. The solution to this problem is not new as it involves the use of Lagrange multipliers, a well-known technique in linear algebra. The next section introduces the methods in use based on Lagrange multipliers. The introduction is in a form that makes understanding of the formulation of the Lagrangian fairly easy to follow.

7.5 The Lagrangian Optimization Function

To present the next sections in clearer light digression from the main SVM, and discussions on Lagrange multipliers are essential. The discussions are based on the excellent treatise on the topic of Lagrangian by Smith [2], which can be used to find the optimal hyperplanes.

The Lagrange multipliers method is a mathematical concept used in solving constrained optimization problems in which the optimization function is differentiable. The objective is to find the minimum or maximum (minmax) of a function $f(x_1, \cdots, x_n)$, which is constrained or restricted in terms of its boundaries of influence by a constraint function. In most cases, the problem encountered by researchers in this area is how to formulate the Lagrangian, the new function that combines the function to be optimized with the constraint function. The Lagrangian is basically an expression which equates the derivative of the function to be optimized with the derivative of the constraint multiplied with the Lagrange multiplier(s). Formally, if the constraint function is $g(x)$, the derivative of the Lagrangian is

$$\nabla f(x) = \lambda \nabla g(x) \qquad (7.26)$$

where λ is the Lagrange multiplier. Equation (7.26) equates the derivative of the function to be optimized with that of the constraint multiplied with the Lagrange multiplier.

7.5.1 Optimization Involving Single Constraint

To illustrate the meaning of the expression (7.26), suppose $f(x)$ and $g(x)$ are vectors, the equation says that the two functions have equal gradients when that of the constraint is multiplied by the Lagrange multiplier. Two vectors will have the same gradient only if they are parallel. Thus, λ in Equation (7.26) is a scaling factor for the two vectors. One vector is a scaled form of the other, where the scaling factor is λ. Therefore, the Lagrangian (the equation which combines the function to be optimized) and the constraint can be written as

$$f(x) = \lambda g(x) \qquad (7.27)$$

Often this expression is written as

$$L(x) = f(x) - \lambda g(x) \qquad (7.27a)$$

This form of the expression further complicates the concept for the uninitiated in the area. The optimization objective therefore becomes that of seeking to find the solution for

$$\nabla L(x, \lambda) = 0 \qquad (7.27b)$$

This equation is identical to Equation (7.26). Three forms of constraints are used in the current literature. These are equality, less than or greater than constraints.

Example: Find the extreme values which
   Maximize $f(x) = 3 - x^2 - 2y^2$
   Subject to constraint: $g(x) = x + 2y - 1 = 0$

Solution: First form the Lagrangian for the optimization problem. This has three variables x, y and λ as

$$L(x, y, \lambda) = f(x) - \lambda g(x) = 3 - x^2 - 2y^2 - \lambda(x + 2y - 1)$$

Note, the Lagrangian combines the optimization function and its constraint(s) into a single expression. To find the extreme values (maximum or minimum), we differentiate the Lagrangian with respect to each of the variables and set them to zero and solve them simultaneously. They are

$$\frac{\partial L(x, y, \lambda)}{\partial x} = 2x - \lambda = 0; \Rightarrow \lambda = 2x$$

$$\frac{\partial L(x, y, \lambda)}{\partial y} = 4y - 2\lambda = 0 \Rightarrow 4y - 4x = 0$$

$$x = y$$

$$\frac{\partial L(x, y, \lambda)}{\partial \lambda} = -(x + 2y - 1) = 0$$

Therefore, from the last equation and with substitution:

$$(x + 2y - 1) = 0$$

$$3x = 1 \Rightarrow x = \frac{1}{3}; \quad y = \frac{1}{3}; \quad \lambda = \frac{2}{3}$$

This means that $f(x) = 3 - x^2 - 2y^2 = 3 - \frac{1}{9} - \frac{2}{9} = 3 - \frac{1}{3}$ and $f(x) = \frac{8}{3}$. The optimum value of the function $f(x)$ subject to the constraint is thus obtained as $\frac{8}{3}$.

7.5.2 Optimization with Multiple Constraints

In many applications, including SVM, the optimization problem requires and contains many constraints. This is similar to the case with one constraint except that several other constraints have been added to the problem. Therefore, the Lagrangian also should reflect the presence of a sum of constraints as in Equation (7.28):

$$L(x, \lambda) = f(x) - \sum_i \lambda_i g_i(x) \qquad (7.28a)$$

In Equation (7.28a), the multiple constraints $g_i(x)$ is a sum of constraints with multiple multiplier λ_i for each constraint. This problem is solved by searching for points where

$$\nabla L(x, \lambda_i) = 0 \qquad (7.28b)$$

This form of Lagrangian is popular in support vector machine applications and demonstrated in this section with an example.

Example 2: Optimize the function

Maximize $f(x, y) = x^2 + y^2$

Subject to constraints $\begin{aligned} g_1(x, y) &= 2x + 1 = 0 \\ g_2(x, y) &= -y + 1 = 0 \end{aligned}$

Solution: We form the Lagrangian first and is

$$\begin{aligned} L(x, y, \lambda) &= f(x, y) - \lambda_1 g_1(x, y) - \lambda_2 g_2(x, y) \\ &= x^2 + y^2 - \lambda_1(2x + 1) - \lambda_2(1 - y) \end{aligned}$$

Since there are two constraints, the Lagrangian also have two multipliers. Next we take the derivative of the Lagrangian with respect to four variables x, y, λ_1 and λ_2, which are:

$$\frac{\partial L(x, y, \lambda)}{\partial x} = 2x - 2\lambda_1 = 0 \Rightarrow x = \lambda_1$$

$$\frac{\partial L(x, y, \lambda)}{\partial y} = 2y + \lambda_2 = 0 \Rightarrow y = \frac{-\lambda_2}{2}$$

$$\frac{\partial L(x, y, \lambda)}{\partial \lambda_1} = -(2x + 1) = 0 \Rightarrow x = -\frac{1}{2}$$

$$\frac{\partial L(x, y, \lambda)}{\partial \lambda_2} = -(1 - y) = 0 \Rightarrow y = 1$$

Therefore, $\lambda_1 = -\frac{1}{2}, \lambda_2 = -2, x = -\frac{1}{2}$ and $y = 1$. The function $f(x, y) = \frac{5}{4}$.

7.5.2.1 Single inequality constraint

Inequality constraints are used rampantly in optimization problems. The Lagrangian solutions are not anything different from the ones we have encountered for the equality constraints above. They are formed the same way the derivatives are taken in the same manner. A further requirement for the inequality constraints is that we must use the following, that when

$$g(x) \geq 0 \Rightarrow \lambda \geq 0$$

$$g(x) \leq 0 \Rightarrow \lambda \leq 0$$

$$g(x) = 0 \Rightarrow \lambda \quad is\,not\,constrained$$

Hence, if the constraints are less than or equal to zero, the multipliers are also less than or equal to zero. If the constraint is greater than or equal to zero, the multiplier is also greater than or equal to zero. Consider the following optimization problem with inequality constraints.

Example 3: Solve

Maximize $f(x, y) = x^3 + 2y^2$
Subject to the constraint $g(x, y) = x^2 - 1 \geq 0$

Solution: The solution to the problem follows the same procedure as for the equality constraint types. We form the Lagrangian function first, which is

$$L(x, y, \lambda) = f(x, y) - \lambda g(x, y) = x^3 + 2y^2 - \lambda(x^2 - 1)$$

The derivatives are

$$\frac{\partial L(x, y, \lambda)}{\partial x} = 3x^2 - 2\lambda x = 0$$

$$\frac{\partial L(x, y, \lambda)}{\partial y} = 4y = 0 \Rightarrow y = 0$$

$$\frac{\partial L(x, y, \lambda)}{\partial \lambda} = x^2 - 1 = 0 \Rightarrow x = \pm 1$$

We can now substitute for x and y in the first derivative to have the solutions as follows.

When $x = 1, y = 0$, we have $3 - 2\lambda = 0 \Rightarrow \lambda = \frac{3}{2}$
When $x = -1, y = 0$, we have $3x^2 - 2\lambda x = 3 + 2\lambda = 0 \Rightarrow \lambda = -\frac{3}{2}$.

The value of the function becomes:

$$f(x, y) = x^3 + 2y^2 = \pm 1.$$

We have two values of λ, one positive and the other negative. Since the constraint is given as greater than inequality, we also require to use the positive value $\lambda = \frac{3}{2}$. Therefore, the Lagrange multiplier is $\lambda = 1.5$.

7.5.2.2 Multiple inequality constraints

The use of multiple constraints and in particular multiple inequality constraints are very popular in SVM applications. They require adherence to certain rules in relation to the values of the Lagrange multipliers involved. The simple rules are that if a greater than inequality constraint is used, the Lagrange multiplier must be equal to or greater than zero. If it is less than inequality, then the Lagrange multiplier should also have value less than or equal to zero. This is stated formally with an example. Given an optimization problem of the form to

Maximize $f(x, y) = x^3 + y^3$

Subject to constraints

$$g_1(x, y) = x^2 - 1 \geq 0$$
$$g_2(x, y) = y^2 - 1 \geq 0$$

The Lagrange multipliers must also be equal to or greater than zero, $\lambda_1 \geq 0$ and $\lambda_2 \geq 0$. However, if the optimization problem were of the form

Maximize $f(x, y) = x^3 + y^3$

Subject to constraints

$$g_1(x, y) = x^2 - 1 \geq 0$$
$$g_2(x, y) = y^2 - 1 \leq 0$$

It is required for this case that the Lagrange multipliers obtained or used be of the form $\lambda_1 \geq 0$ and $\lambda_2 \leq 0$, respectively.

7.5.3 Karush–Kuhn–Tucker Conditions

Five constraint conditions which must be obeyed in SVM analysis were stated by Karush, Kuhn and Tucker (KKT). They relate the derivatives of Lagrangians to the constraints that need to be fulfilled in the solutions. They are stated without proof as

$$
\left.
\begin{aligned}
\frac{\partial L(\vec{w}, b, \lambda)}{\partial \vec{w}} &= \vec{w} - \sum_i \lambda_i y_i \vec{x}_i = 0 \quad &(i) \\
\frac{\partial L(\vec{w}, b, \lambda)}{\partial b} &= -\sum_i \lambda_i y_i = 0 \quad &(ii) \\
y_i(\vec{w}^T \cdot \vec{x} + b) - 1 &\geq 0 \quad &(iii) \\
\lambda_i &\geq 0 \quad &(iv) \\
\lambda_i(y_i(\vec{w} \cdot \vec{x} + b) - 1) &= 0 \quad &(v)
\end{aligned}
\right\} \quad (7.29)
$$

In Equations (7.29), the range of i is from 1 to m. Constraints (i) and (ii) are called the stationary constraints. At a stationary point, the function stops increasing or decreasing. Constraint (iii) is the primal feasibility condition.

The dual feasibility condition is given by constraint (iv) and constraint (v) is the complementary slackness condition.

In the sequel, we now apply the discussions so far on hyperplanes and optimization techniques in the analysis of support vector machines.

7.6 SVM Optimization Problems

We will now return to the solution of the optimization problem posed in Equations (7.24) and (7.25). Two optimization schemes are used for the analysis of SVM data in this chapter. They are the primal and the dual schemes. They are discussed separately. They are discussed first for linearly separable data and later for nonlinear data, which require the use of slack variables.

7.6.1 The Primal SVM Optimization Problem

For the primal SVM optimization, we introduce Lagrange multipliers α_i. This is associated with minimization of $L_p(w, b, \alpha)$, the primal optimizer given by the Lagrangian

$$\text{Minimize } L_p(w, b, \alpha) = \frac{\|w\|^2}{2} - \sum_{i=1}^{N} \alpha_i(y_i(w^T \cdot x_i + b) - 1) \tag{7.30}$$

Subject to the constraint: $\alpha_i \geq 0$

The optimization is solved by taking derivatives of the $L_p(w, b, \alpha)$ with respect to w and also with respect to b and setting them to zero. In the first case

$$\frac{\partial}{\partial w} L_p(w, b, \alpha) = 0 \Rightarrow w - \sum_{i=1}^{N} \alpha_i x_i y_i = 0 \tag{7.31}$$

Therefore

$$w = \sum_{i=1}^{N} \alpha_i x_i y_i \tag{7.32}$$

Therefore, we can solve for the weights based on the data vectors and the Lagrangian multipliers.

The second partial derivative with respect to b provides the means for estimating the values of the Lagrange multipliers using the expression:

$$\frac{\partial}{\partial b} L_p(\boldsymbol{w}, b, \alpha) = 0 = \sum_{i=1}^{N} \alpha_i y_i \tag{7.33}$$

By setting this to zero, we have

$$\sum_{i=1}^{N} \alpha_i y_i = 0 \tag{7.34}$$

Equations (7.32) and (7.34) are the solutions that are used to determine the Lagrange multipliers and the weights $\boldsymbol{w}$. The product of the Lagrange multipliers and the class indices must sum to zero over the classes. In the training leading to optimization, support vectors are the points for which $\alpha_i \neq 0$. The vectors lie on the margin and satisfy the expression

$$y_i(\mathbf{x_i} \cdot \mathbf{w}^{\mathrm{T}} + b) - 1 = 0 \quad \forall i \in S$$

S contains the indices to the support vectors. For classification purposes, vectors for which $\alpha_i = 0$ are irrelevant.

7.6.2 The Dual Optimization Problem

In this section, we extend the primal problem to the dual problem in which Lagrangian multipliers α_i are introduced. In the **dual problem**, once $\boldsymbol{w}$ is known, then we know all α_i. Also when we know all the α_i, we also know $\boldsymbol{w}$. This is called the dual problem. The section also covers **soft margins** and use of **kernel functions**. The dual problem is used to convert the primal SVM form into more easily solvable forms. The dual optimization starts with the primal Equation (7.30) by transforming it into a more solvable problem. The dual problem is given by the Equation (7.35)

$$\max_{\alpha} W(\alpha) = \max_{\alpha}(\min_{\alpha} L(\boldsymbol{w}, b, \alpha)) \tag{7.35}$$

In the dual problem, we maximize an objective function

$$\text{Maximize } L_d(\alpha) = \sum_{i=1}^{N} \alpha_i - \frac{1}{2} \sum_{i=1}^{N} \sum_{i,j=1}^{N} \alpha_i \alpha_j y_i y_j \vec{x}_i \cdot \vec{x}_j \tag{7.36a}$$

$$\text{Subject to the constraints } \alpha_i \geq 0 \text{ and } \sum_{i=0}^{N} \alpha_i y_i = 0 \tag{7.36b}$$

The sequence of Lagrangian multipliers $\{\alpha_1, \alpha_2, \cdots, \alpha_N\}$ are obtained one for each point x_i.

The quadratic programming (QP) solution problem specified above has solution. It uses a quadratic objective function subject to linear constraints. They may be solved using greedy algorithms. In the example given in this section, the algorithm defines w in terms of the Lagrangian:

$$w = \sum_{i=1}^{N} \alpha_i y_i x_i \tag{7.37}$$

$$b = y_k - \mathbf{w}^T \mathbf{x}_k \tag{7.38}$$

For all $\mathbf{x}_k, \alpha_k \neq 0$. The solution implies that for $\alpha_i \neq 0$, the corresponding value x_i is a support vector. There is always a global solution to the Lagrangian, which is obtained from the above equation. Notice from Equation (7.36) that we could have actually formulated the dual problem as the dot product of two vectors $\vec{w}_i$ and $\vec{w}_j$, where

$$L_d(\alpha) = \sum_{i=1}^{N} \alpha_i - \frac{1}{2} \sum_{i,j=1}^{N} w_i^T w_j \tag{7.39}$$

The solution to Equation (7.39) can be shown to be:

$$f(\vec{x}) = sign \left(\sum_{i=1}^{N} \alpha_i y_i \vec{x}_i \cdot \vec{x} \right) \tag{7.40}$$

The solution in Equation (7.40) shows that for classification, there is no need to access all the original data but just the dot products and the number of support vectors provides the bound on the number of free parameters.

7.6.2.1 Reformulation of the dual algorithm

The dual formulation requires that we minimize an objective function defined by a quadratic expression subject to a linear constraint. They are given as:

Minimize $\frac{1}{2} \sum_{i=1}^{N} w_i^2$

Subject to $y_i(\vec{w} \cdot \vec{x}_i + b) \geq 1 \quad for\ i = 1, \cdots, N$

As usual, the method of using Lagrange multipliers is adopted. The Lagrange multipliers are defined by the equation

$$L_d(\vec{w}, b, \vec{\alpha}) = \frac{1}{2} \sum_{i=1}^{N} w_i^2 - \sum_{i=1}^{N} \alpha_i(y_i(\vec{w} \cdot \vec{x}_i + b) - 1) \tag{7.41}$$

The vectors $\vec{w}$ and $\vec{\alpha}$ have N elements. The Lagrangian needs to be minimized with respect to b and $\vec{w}$. Also the derivative of the Lagrangian with respect to $\vec{\alpha}$ should vanish subject to the constraint that the value of $\bar{\alpha}_i$ be equal to or greater than zero. The dual problem is

$$L_d(\vec{w}, b, \vec{\alpha}) = \sum_{i=1}^{N} \alpha_i - \frac{1}{2} \sum_{i=1}^{N} \sum_{j=1}^{N} \alpha_i \alpha_j y_i y_j \vec{x}_i \cdot \vec{x}_j \qquad (7.42)$$

The solution to the problem is to maximize

$$\max_{\alpha} L_d(\vec{w}, b, \vec{\alpha}) = \max_{\alpha} \left[\sum_{i=1}^{N} \alpha_i - \frac{1}{2} \sum_{i=1}^{N} \sum_{j=1}^{N} \alpha_i \alpha_j y_i y_j \vec{x}_i \cdot \vec{x}_j \right] \qquad (7.42a)$$

The following derivatives are equated to zero to give the solution earmarked in Equation (7.42):

$$\frac{\partial}{\partial w} L_d(\vec{w}, b, \vec{\alpha}) = 0 \Rightarrow \vec{w} = \sum_{i=1}^{N} \alpha_i y_i \vec{x}_i \qquad (7.42b)$$

$$\frac{\partial}{\partial b} L_d(\vec{w}, b, \vec{\alpha}) = 0 \Rightarrow \sum_{i=1}^{N} \alpha_i y_i \qquad (7.42c)$$

The values $\alpha_i \geq 0$; $i = 1, 2, \cdots, N$. Hence, from Equations (7.42b) and (7.42c), the optimal solutions to the dual problem are:

$$\bar{w} = \sum_{i=1}^{N} \bar{\alpha}_i y_i \vec{x}_i \qquad (7.42d)$$

$$\bar{b} = -\frac{1}{2} \bar{w} \cdot [\vec{x}_i + \vec{x}_j]$$

where $\vec{x}_i$ and $\vec{x}_j$ are any support vectors that satisfy $\bar{\alpha}_i, \bar{\alpha}_j > 0$; $y_i = 1$; $y_j = -1$. There is always a global maximum for the Lagrangian. The hard classifier therefore is

$$f(\boldsymbol{x}) = sign(\mathbf{\bar{w}}^T \cdot \mathbf{\vec{x}} + \bar{b}) \qquad (7.43)$$

The soft classifier is also given by the expression

$$f(\boldsymbol{x}) = h(\mathbf{\bar{w}} \cdot \mathbf{\vec{x}} + \bar{b}) \qquad (7.44)$$

$$h(x) = \begin{cases} +1 & x > 1 \\ x & -1 \leq x \leq 1 \\ -1 & x < -1 \end{cases} \qquad (7.45)$$

The soft classifier includes the possibility of a data point lying within the margin region where there is no training data and with classification value $-1 \leq x \leq 1$ [3]. It is useful when the classifier is tested within the margin region.

Example 4: Consider the following points:

$$\begin{aligned}
\mathbf{x_1} &= [0.0\ 0.0]^T & y_1 &= -1 \\
\mathbf{x_2} &= [1.0\ 0.0]^T & y_2 &= +1 \\
\mathbf{x_3} &= [0.0\ 1.0]^T & y_3 &= +1
\end{aligned}$$

The three points lie on the origin of the Cartesian axes, on the x-axis and on the y-axis. Hence $N = 3$. The optimization problem is to

$$\text{Maximize } L(\alpha) = \sum_{i=1}^{3} \alpha_i - \frac{1}{2} \sum_{i=1}^{3} \sum_{j=1}^{3} \alpha_i \alpha_j y_i y_j (\mathbf{x_i} \cdot \mathbf{x_j})$$

$$\alpha_i \geq 0, \quad i = 1, \ldots, 3$$

Subject to the constraints
$$\sum_{i=1}^{3} \alpha_i y_i = 0$$

To include the effect of the second constraint, it is necessary to introduce a second multiplier, which leads to the Lagrangian function:

$$\begin{aligned}
L_d(\alpha, \lambda) &= \sum_{i=1}^{3} \alpha_i - \frac{1}{2} \sum_{i=1}^{3} \sum_{j=1}^{3} \alpha_i \alpha_j y_i y_j (\mathbf{x_i} \cdot \mathbf{x_j}) - \lambda \sum_{i=1}^{3} \alpha_i y_i \\
&= (\alpha_1 + \alpha_2 + \alpha_3) - \frac{1}{2}\alpha_2^2 - \frac{1}{2}\alpha_3^2 - \lambda(\alpha_1 y_1 + \alpha_2 y_2 + \alpha_3 y_3) \\
&= (\alpha_1 + \alpha_2 + \alpha_3) - \frac{1}{2}\alpha_2^2 - \frac{1}{2}\alpha_3^2 - \lambda(-\alpha_1 + \alpha_2 + \alpha_3)
\end{aligned}$$

To solve this problem, we need to differentiate the new Lagrangian with respect to the variables α_i and λ and set each differential to zero $\alpha_1 = 4$, $\alpha_2 = 2$, $\alpha_3 = 2$, $\lambda = -1$. Therefore, we can now solve for the weights using

$$\mathbf{w} = \sum_{i=1}^{3} \alpha_i y_i \mathbf{x_i} = [2\ 2]^T \quad \text{and} \quad b = 1 - \mathbf{w}^T \mathbf{x}$$

The equation for the optimal hyperplane will next have to be used to determine the margin.

7.7 Linear SVM (Non-linearly Separable) Data

We have so far assumed that data can be mostly classified linearly and that they are separated well using hyperplane. This assumption is flawed as it is possible and a natural occurrence that data of one time could take the form of data of another form. While this is not the norm, the possibility is always there due to noise in measurements or noise in the system itself. There also could be outliers thereby making such data non-linear. When data points mix, the probability of misclassification increases. What are the techniques for classifying such data?

Two of the popular techniques for separating such data include the use of the so-called "slack variable" and the use of "kernel". Both of them are discussed with analysis in this section.

7.7.1 Slack Variables

Slack variables are used to indicate the level of error in classification of a data sample from the separating hyperplane. Each data sample is assigned a slack variable, which defines its level of classification.

The slack variables are positive numbers including zero ($\xi_i \geq 0$). A sample that is classified without error has zero slack variable. In Figure 7.9, the sample with blue circle close to the hexagonal class is going to be wrongly classified because it lies in the separating hyperplane close to the hexagonal class. Also the hexagonal sample in the middle of the separating hyperplanes could also be wrongly classified.

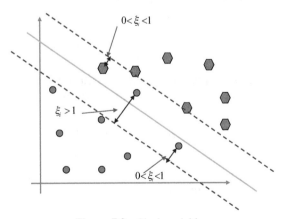

Figure 7.9 Slack variables.

7.7.1.1 Primal formulation including slack variable

For the primal formulation inclusive of slack variables, we seek to minimize for the soft-margin linear SVM case the objective function

$$\text{Minimize} \quad \phi_p(w) = \frac{1}{2}\sum_{i=1}^{N} w_i^2 + C\sum_{i=1}^{N} \xi_i \tag{7.45a}$$

$$\text{Subject to} \quad y_i(\vec{w}\cdot\vec{x}_i + b) \geq 1 - \xi \quad for \ i = 1,\cdots,N \tag{7.45b}$$

Therefore, when a new data sample x is to be classified, the classifier is given by the function $f(x) = sign(\vec{w}\cdot\vec{x}+b)$. The constant C>0 in the slack variable equation is a penalty on the summation term $\sum_{i=1}^{N} \xi_i$. It prevents and controls overfitting and therefore acts as a tradeoff between the training error and the margin. The solution to Equation (7.45a) subject to Equation (7.45b) is given by the saddle point of the following equation

$$L_p(\boldsymbol{w}, b, \alpha, \xi, \beta) = \frac{1}{2}\|\boldsymbol{w}\|^2 + C\sum_{i=1}^{N} \xi_i$$

$$- \sum_{i=1}^{N} \alpha_i \left[(\boldsymbol{w}^T \cdot x + b)y_i + \xi_i - 1 \right] - \sum_{i=1}^{N} \beta_i\xi_i$$

where α_i, β_i are the Lagrange multipliers. The Lagrangian is minimized with respect to $\boldsymbol{w}, \xi, b$. It is also minimized with respect to $\alpha_i, \beta_i \geq 0$.

7.7.1.2 Dual formulation including slack variable

Similar to the primal formulation, the dual formulation is derived from the primal formulation. It is specified by

$$\max_{\alpha,\beta} W\left(\alpha, \beta\right) = \max_{\alpha,\beta} \left[\min_{w,b,\xi} L\left(\boldsymbol{w}, b, \alpha, \beta, \xi\right) \right] \tag{7.46}$$

The minimum of this expression is obtained through the solution of the partial derivatives of L with respect to w, b and ξ, where

$$\frac{\partial}{\partial b} L_d(\boldsymbol{w}, b, \alpha) = 0 \Rightarrow \sum_{i=1}^{N} \alpha_i y_i \tag{7.47}$$

$$\frac{\partial}{\partial \boldsymbol{w}} L_d(\boldsymbol{w}, b, \alpha) = 0 \Rightarrow \boldsymbol{w} = \sum_{i=1}^{N} \alpha_i y_i \boldsymbol{x_i} \tag{7.48a}$$

$$\frac{\partial}{\partial \xi_i} L_d(\boldsymbol{w}, b, \alpha) = 0 \Rightarrow \alpha_i + \beta_i = C \tag{7.48b}$$

Therefore, the objective function and the slack variable equation (constraint) for the dual problem are given by the expressions:

$$\max_{\alpha} W(\alpha) = \sum_{i=1}^{N} \alpha_i - \frac{1}{2} \sum_{i=1}^{N} \sum_{j=1}^{N} \alpha_i \alpha_j y_i y_j \vec{x}_i \cdot \vec{x}_j; \ for \ i = 1, \cdots, N$$

(7.49a)

The solution is then

$$\bar{\alpha} = -\sum_{i=1}^{N} \alpha_i + \frac{1}{2} \sum_{i=1}^{N} \sum_{j=1}^{N} \alpha_i \alpha_j y_i y_j \vec{x}_i \cdot \vec{x}_j \qquad (7.50)$$

$$\text{With constraint } 0 \leq \alpha_i \leq C \, and \, \sum_{i=1}^{N} \alpha_i y_i = 0 \qquad (7.50a)$$

The constant C>0 provides the bound for the Lagrangian parameters α_i. The value of w is found by using the expression:

$$w = \sum_{i=1}^{N} \alpha_i y_i x_i \qquad (7.51)$$

The value of b is obtained from the solution of the equation $\alpha_i(y_i(\vec{w}^T \cdot \vec{x}_i + b) - 1) = 0$.

Three possibilities are considered for the Lagrangian multipliers. When

$$\left.\begin{array}{l} \alpha_i = 0 \Rightarrow y_i \left(w^T \cdot x_i + b \right) > 1 \\ 0 < \alpha_i < C \Rightarrow y_i \left(w^T \cdot x_i + b \right) = 1 \\ \alpha_i = C \Rightarrow y_i \left(w^T \cdot x_i + b \right) < 1 \\ (for \ samples \ with \ \xi_i > 0) \end{array}\right\}$$

(7.52)

The three cases provide the bounds for the hyperplanes. In the next section, discussions on how to choose values of C an area of intense research in SVM will be discussed.

7.7.1.3 Choosing C in soft margin cases

The role of C is major in preventing overfitting. Very small values of C normally lead to misclassification. Small values of C lead to training errors due to training samples that are not in appropriate positions. Consequently, classification performance is reduced. When C is very small underfitting is usually the result. We thus end up having naïve classifiers. The choice of C is therefore crucial for having a good solution to the problem. As C increases to infinity, the solution also converges to the optimal separating hyperplane solution.

7.7.2 Non-linear Data Classification Using Kernels

The use of soft margins allows some errors to be made in the estimation of the SVM margins. Also kernel functions allow introduction of nonlinearity into the SVM solutions. Transformation has played a significant role in the solution of algebraic functions, in signal processing and statistical analysis. In these fields, data or signals in a domain where solutions become difficult are transformed into alternate solutions, which makes their solutions a lot easier. This gave birth to the Laplace transform, Fourier transform, orthogonal transforms, wavelet transforms, principal component analysis and a lot of other transform methods. "Kernel tricks" as used in SVM represent adoption of similar concept to transform non-linearly separable data in the input space using a kernel to a feature space where such data become separated and lead to efficient classification. A kernel is a mapping function which transforms data from the input space to a more solvable feature space. The class of separable functions is very popular in SVM.

Consider a data set mapped in the feature space which does not yield to linear separation in the input space using hyperplanes directly. The idea behind the use of a kernel function is to facilitate the operations to be performed in the input space instead of the higher dimensional feature space. Input space is defined as the space where the points x_i are located. The feature space is also defined as the space in which the transform point $\phi(x)$ lies after transformation. After transformation, the inner product is performed in the input space instead of the feature space. For this to be useful, it requires a large training set. Hence, the computation is a function of the number of training points. For illustration, consider the example given in the scatter grams of Figure 7.10. A kernel (function) is used to transform the data points from the original feature space to another feature space in which the points are clearly separable.

We will start from the well-known case and represent the linear classifier with a linear kernel

$$K(x_i, x_j) = x_i^T . x_j \tag{7.53}$$

For the nonlinear case, we will represent the kernel as a mapping of the data point x to a higher-dimensional space through the use of a function Φ, where

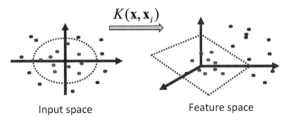

$$K(\mathbf{x}, \mathbf{x}_i)$$

Input space   Feature space

Figure 7.10 Transformation from original feature space to higher-dimensional feature space.

the transformation is

$$\Phi : x \rightarrow \phi(x)$$

In the feature space, the inner product has an equivalent kernel in the input space, which is often represented as

$$K(x, y) = k(\boldsymbol{x}) \cdot k(\boldsymbol{y}) \tag{7.54}$$

$K(x, y)$ must possess defined characteristics and they must satisfy the Mercer conditions, which are

$$K(\mathbf{x}, \mathbf{y}) = \sum_{m=1}^{\infty} \alpha_m \psi(\boldsymbol{x}) \psi(\boldsymbol{y}), \quad \alpha_m \geq 0 \tag{7.55}$$

Therefore, a function $K(x, y)$ is a kernel if there exists transformation $\phi(x)$ such that $K(x_i, x_j) = \phi(x_i)^T \cdot \phi(x_j)$. Equivalently, a Kernel matrix is symmetric positive semi-definite. A second statement of the Mercer condition is that $K(x_i, x_j)$ is a kernel for a function $g(\mathbf{x})$ and $g(\mathbf{y})$ such that $\int g^2(\mathbf{x}) d\boldsymbol{x} < \infty$ (is finite) and

$$\iint K(\mathbf{x}, \mathbf{y}) g(\mathbf{x}) g(\mathbf{y}) d\mathbf{x} d\mathbf{y} > 0 \tag{7.56}$$

The kernel function which satisfies the Mercer conditions using the data points is formally defined as

$$K(x_i, x_j) = \phi(x_i)^T . \phi(x_j) \tag{7.57}$$

The kernel is an inner product of the functions of the data points. The kernel is used to construct an optimal separator in the higher dimension space. Some examples of kernel functions that satisfy the Mercer conditions are polynomial, Gaussian radial basis function, perceptron, exponential radial basis, splines, Fourier kernels series and Bsplines.

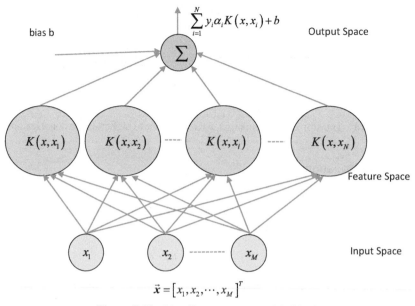

Figure 7.11   Non-linear none-separable SVM.

In the non-linear case being discussed, the training data X are not linearly separable. The kernel function maps the data into a feature space from an input space where they are linearly separable (Figure 7.11). In Figure 7.11, all possible nodes are shown in the hidden layer at the feature space. In practice, assuming there are p support vectors, then the hidden layer is evaluated with p hidden nodes instead of N nodes. The support vector set is thus $S = [x_1, x_2, \cdots, x_p]^T = [s_1, s_2, \cdots, s_p]^T$. In general, we can compose the kernel matrix such that

$$
K(x, x_i) = \begin{bmatrix} k(\mathbf{x}_1, \mathbf{x}_1) & k(\mathbf{x}_1, \mathbf{x}_2) & ... & k(\mathbf{x}_1, \mathbf{x}_m) \\ k(\mathbf{x}_2, \mathbf{x}_1) & k(\mathbf{x}_2, \mathbf{x}_2) & ... & k(\mathbf{x}_2, \mathbf{x}_m) \\ ... & ... & ... & ... \\ k(\mathbf{x}_m, \mathbf{x}_1) & k(\mathbf{x}_m, \mathbf{x}_2) & ... & k(\mathbf{x}_m, \mathbf{x}_m) \end{bmatrix} \tag{7.58}
$$

Good kernels lead to symmetric positive-definite matrices, which is very useful in schemes that depend on the use of kernel matrices for learning or the so-called KM-learning. Kernels provide a measure of similarity between data. Hence, the optimization problem reduces to

$$\text{Maximize}: W(\alpha) = \sum_{i=1}^{N} \alpha_i - \frac{1}{2} \sum_{i=1}^{N} \sum_{j=1}^{N} \alpha_i \alpha_j y_i y_j K(\mathbf{x}_i, \mathbf{x}_j)$$

(7.59)

$$\text{Subject to } \alpha_i \geq 0, \quad i = 1, \ldots, N, \quad \text{and} \quad \sum_{i=1}^{N} \alpha_i y_i = 0$$

The kernel function is used in Equation (7.59) in place of the dot product of the vectors. The decision function for the non-linear SVM is given by the expression

$$f(\mathbf{x}) = sgn\left(\sum_{i=1}^{N} y_i \alpha_i K(\mathbf{x}, \mathbf{x}_i) + b \right)$$

(7.60)

The decision function (Equation (7.60)) is equally modified with the kernel taking the place of the dot product of the vectors.

Observe how similar SVM with kernels as shown in Figure 7.11 is similar to neural networks with a hidden layer. In the next sections, these kernel functions are introduced. We also briefly cover summations and product of kernel functions. Kernels are discussed in the next section.

7.7.2.1 Polynomial kernel function
Consider the polynomial kernel function

$$K(x_i, x_j) = (1 + \mathbf{x}_i^T \cdot \mathbf{x}_j)^2 = 1 + 2(\mathbf{x}_i^T \cdot \mathbf{x}_j) + (\mathbf{x}_i^T \cdot \mathbf{x}_j)^2$$

$$= 1 + 2(\mathbf{x}_{i1}, \mathbf{x}_{i2}) \cdot \begin{pmatrix} x_{j1} \\ x_{j2} \end{pmatrix} + \left[(\mathbf{x}_{i1}, \mathbf{x}_{i2}) \cdot \begin{pmatrix} x_{j1} \\ x_{j2} \end{pmatrix} \right]^2$$

(7.61)

We can show that the kernel function can be written as an inner product of two functions of the form $\phi(x_i)^T \cdot \phi(x_j)$. This inner product is demonstrated by expanding and factoring the polynomial kernel as where the polynomial kernel is:

$$K(x_i, x_j) = (1 + \mathbf{x}_i^T \cdot \mathbf{x}_j)^2 = 1 + 2(\mathbf{x}_i^T \cdot \mathbf{x}_j) + (\mathbf{x}_i^T \cdot \mathbf{x}_j)^2$$

$$= 1 + 2(\mathbf{x}_{i1}, \mathbf{x}_{i2}) \cdot \begin{pmatrix} x_{j1} \\ x_{j2} \end{pmatrix} + \left[(\mathbf{x}_{i1}, \mathbf{x}_{i2}) \cdot \begin{pmatrix} x_{j1} \\ x_{j2} \end{pmatrix} \right]^2$$

(7.62)

By expanding and factoring this expression further, we have

$$K(x_i, x_j) = 1 + 2(\boldsymbol{x}_{i1}, \boldsymbol{x}_{i2}) \cdot \begin{pmatrix} \boldsymbol{x}_{j1} \\ \boldsymbol{x}_{j2} \end{pmatrix} + \left[(\boldsymbol{x}_{i1}, \boldsymbol{x}_{i2}) \cdot \begin{pmatrix} \boldsymbol{x}_{j1} \\ \boldsymbol{x}_{j2} \end{pmatrix} \right]^2$$

$$= 1 + 2[\boldsymbol{x}_{i1}\boldsymbol{x}_{j1} + \boldsymbol{x}_{i2}\boldsymbol{x}_{j2}] + [\boldsymbol{x}_{i1}\boldsymbol{x}_{j1} + \boldsymbol{x}_{i2}\boldsymbol{x}_{j2}]^2$$

$$= 1 + 2\boldsymbol{x}_{i1}\boldsymbol{x}_{j1} + 2\boldsymbol{x}_{i2}\boldsymbol{x}_{j2} + (\boldsymbol{x}_{i1}\boldsymbol{x}_{j1})^2 + (\boldsymbol{x}_{i2}\boldsymbol{x}_{j2})^2 \qquad (7.63)$$

$$+ 2\boldsymbol{x}_{i1}\boldsymbol{x}_{j1}\boldsymbol{x}_{i2}\boldsymbol{x}_{j2}$$

$$= [1\,\boldsymbol{x}_{i1}^2\,\sqrt{2}\boldsymbol{x}_{i1}\boldsymbol{x}_{i2}\,\boldsymbol{x}_{i2}^2\,\sqrt{2}\boldsymbol{x}_{i1}\,\sqrt{2}\boldsymbol{x}_{i2}]^T$$

$$[1\,\boldsymbol{x}_{j1}^2\,\sqrt{2}\boldsymbol{x}_{j1}\boldsymbol{x}_{j2}\,\boldsymbol{x}_{j2}^2\,\sqrt{2}\boldsymbol{x}_{j1}\,\sqrt{2}\boldsymbol{x}_{j2}]$$

Therefore, the kernel function is

$$K(x_i, x_j) = \phi(\boldsymbol{x}_i)^T \cdot \phi(\boldsymbol{x}_j)$$

$$\phi(\boldsymbol{x}_i) = \begin{bmatrix} 1 & \boldsymbol{x}_{i1}^2 & \sqrt{2}\boldsymbol{x}_{i1}\boldsymbol{x}_{i2} & \boldsymbol{x}_{i2}^2 & \sqrt{2}\boldsymbol{x}_{i1} & \sqrt{2}\boldsymbol{x}_{i2} \end{bmatrix} \qquad (7.64)$$

$$\phi(\boldsymbol{x}_j) = \begin{bmatrix} 1 & \boldsymbol{x}_{j1}^2 & \sqrt{2}\boldsymbol{x}_{j1}\boldsymbol{x}_{j2} & \boldsymbol{x}_{j2}^2 & \sqrt{2}\boldsymbol{x}_{j1} & \sqrt{2}\boldsymbol{x}_{j2} \end{bmatrix}$$

7.7.2.2 Multi-layer perceptron (Sigmoidal) kernel

The sigmoidal kernel is used also in neural networks as an activation function. In that regard, when it is used in SVM, there is commonality between the two. For such cases, the SVM is equivalent to a two-layer feed forward neural network. The sigmoidal kernel is

$$K(\mathbf{x}_i, \mathbf{x}_j) = \tanh(k\mathbf{x}_i^T \cdot \mathbf{x}_j - \delta) \qquad (7.65)$$

where k is the scale and δ is the offset. For some values of k and, δ it is not a kernel. This is due to the nature of the $\tanh(x)$ function.

7.7.2.3 Gaussian radial basis function

Another example of a kernel is the Gaussian Radial Basis Function (GRBF) given by the function

$$K(x_i, x_j) = \exp\left(-\frac{1}{2\sigma^2} \|x_i - x_j\|^2 \right) \qquad (7.66)$$

It is used to lift data from a low dimension to a much higher dimension where they are separable. The kernel depends on the statistics of the data set. In some applications, finding first the mean and deviations of the data set is helpful in allowing several GRBFs to be used with each one centered on the group statistics.

A variant of the GRBF is the Exponential Radial Basis Function given by a similar expression, which is

$$K\left(x_i, x_j\right) = \exp\left(-\frac{1}{2\sigma^2}|x_i - x_j|\right) \qquad (7.67)$$

It produces discontinuous results and is useful for data sets with discontinuities.

7.7.2.4 Creating new kernels

New kernels can be created from well-known kernels by using a host of tricks including scaling, superposition, products of kernels and symmetry [4]. Examples of such tricks are listed in the following equations for creating new kernels.

$$K(\mathbf{x}, \mathbf{y}) = \beta \cdot K_1(\mathbf{x}, \mathbf{y})$$

$$K(\mathbf{x}, \mathbf{y}) = K_1(\mathbf{x}, \mathbf{y}) \cdot K_2(\mathbf{x}, \mathbf{y})$$

$$K(\mathbf{x}, \mathbf{y}) = \lambda K_1(\mathbf{x}, \mathbf{y}) + (1 - \lambda)K_2(\mathbf{x}, \mathbf{y})$$

$$K(\mathbf{x}, \mathbf{y}) = \frac{K_1(\mathbf{x}, \mathbf{y})}{\sqrt{K_1(\mathbf{x}, \mathbf{x})}\sqrt{K_1(\mathbf{y}, \mathbf{y})}} - cosine\ similarity \qquad (7.68)$$

$$K(\mathbf{x}, \mathbf{y}) = f(\mathbf{x}) \cdot f(\mathbf{y})$$

$$K(\mathbf{x}, \mathbf{y}) = K_3(\phi(\mathbf{x}), \phi(\mathbf{y}))$$

$$K(\mathbf{x}, \mathbf{y}) = \mathbf{x}'P\mathbf{y} \quad symmetric\ definite\ positive$$

$f(x)$ is real-valued, $\beta > 0$ and $0 \le \lambda \le 1$.

References

[1] Alexander Statnikov, Douglas Hardin, Isabelle Guyon, Constantin F. Aliferis, "A Gentle Introduction to Support Vector Machines in Biomedicine", in Biomedical and Health Informatics: From Foundations to Applications and Policy, San Francisco, USA, Nov. 14–18, 2009, pp. 1–207.

[2] Baxter Tyson Smith, "Lagrange Multipliers Tutorial in the Context of Support Vector Machines", Memorial University of Newfoundland, Canada, pp. 1–21, 2004.

[3] Steve Gunn, "Support Vector Machines for Classification and Regression", ISIS Report, May 14, 1998, University of Southampton, United Kingdom.

[4] Jean-Michel Renders "Kernel Methods in Natural Language Processing", Xerox Research Center Europe (France), ACL'04 TUTORIAL.

8

Artificial Neural Networks

8.1 Introduction

A neural network is fundamentally an artificial calculating machine which mimics how cells (neurons) in the human body communicate. It takes inputs from a user and uses its internal functions to make calculations, output a result and takes decisions based on its internal structure. In its simplest form, a neural network is a neuron. While it is semantically not correct to call a neuron a network, it is however the basic template for a neural network. Consider the neuron in Figure 8.1. It has several inputs and produces an output. For now, the internal structure of a neuron remains hidden but will be revealed eventually.

8.2 Neuron

The neuron may be viewed as a single computing site inside a typical neural network. Its algorithmic details on how it computes and how it outputs its results are yet to be specified. Suffice it however to say that it has the fundamental building block of a thinking machine at its atomic level. One of the basic functions of the neuron is to provide a weighted superposition of its inputs. The weights are indicated in Figure 8.1 on the input links or (sinapses) to the brain of the neuron. Sinapses take numbers and multiply them with weights. Neurons sum the results from sinapses and perform activations on them before outputting results into sinapses. The weighted sum undertaken by neurons can be written as in Equation (8.1):

$$S = \sum_{i=1}^{N} \omega_i x_i \qquad (8.1)$$

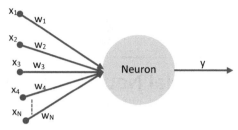

Figure 8.1   A neuron.

Observe that this sum is a linear combination of inputs. The weights may be viewed as the level of importance the neuron attaches to each of the input signals to it. Should the neuron keep on firing irrespective of what the result of the combination of inputs is? In decision making, doing so is a bad idea. There should be a point at which the neuron should issue a result and that point is often given by the so-called activation function. An activation function is used as the decision machine in the neuron to tell it when to issue an output signal. The neuron uses the activation function to undertake further processing on the linear weighted sum S before it can issue a result y or fire.

Consider an athlete who like any other athlete is susceptible to injury and aging. They affect not only the performance of the athlete but when the athlete retires or calls it quicks from his beloved sports.

The inputs to the decision-making machine when the athlete decides to retire could be a weighted sum of several variables including age, injury, recovery and illnesses. Let the weights attached by a club to these inputs be as follows: age (1), injury (5), recovery (–2), illness (3). The following table contains a scenario which requires that the club makes decision on if they should ask the athlete to retire. Without an activation function, the weighted output from the neuron is

$$y = 1 \times \text{age} + 5 \times \text{injury} - 2 \times \text{recovery} + 3 \times \text{illness}$$

In other words, the club puts more weight on injury compared with illnesses. In Table 8.1 is a typical scenario with the linear combination of inputs and the output value are given.

How should the club decide on whom to ask to retire based on the output value in Table 8.1? One approach is for the club to set a threshold above which the neuron should issue a hint to the club to ask the athlete to consider retiring. This could be done in a rather simple form by adding a negative bias

Table 8.1 Linear data combination and bias at neurons

1	5	−2	3		b = −10
Age	Injury	Recovery	Illness	Output value	y + b
4	3	3	3	22	12
2	1	2	2	11	1
1	2	2	4	19	9
5	1	3	1	7	−3
3	1	1	2	12	2

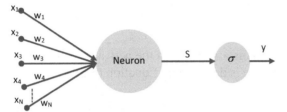

Figure 8.2 Neuron with activation function.

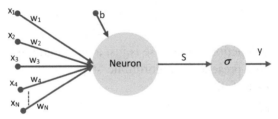

Figure 8.3 Neuron with Bias added.

to the output of the neuron. This is shown as

$$y = 1 \times \text{age} + 5 \times \text{injury} - 2 \times \text{recovery} + 3 \times \text{illness} + b$$

when the bias is set, for example, as −10, one of the athletes has output which is negative, meaning that the athlete is not ready to be advised to retire. One athlete's output result is '2' and is getting close to being asked to retire. Clearly two athletes with y + b = 12 and 9 could be advised to retire even though one of them have an age weight of 1!

To build the idea of when the neuron should activate or fire, an activation function is connected to the output of the neuron as in Figure 8.2.

In clearer terms, each neuron also has a bias input.

The neuron therefore performs the action of summing the product of the input multiplied by the weights. Figure 8.4 is also referred to as a single

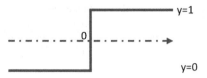

Figure 8.4   Activation function.

perceptron. It perceives its input and issues a response to the input.

$$S = \sum_{i=1}^{N} w_i x_i + b \qquad (8.2)$$

The neuron outputs a value from the activation function in proportion to the value S. This role is depicted by Equation (8.3)

$$f(S) = y = \begin{cases} 1 & if \ \sum_{i=1}^{N} w_i x_i + b \geq 0 \\ 0 & if \ \sum_{i=1}^{N} w_i x_i + b \leq 0 \end{cases} \qquad (8.3)$$

The activation function in this case is just a threshold function (Figure 8.4).

We can now present the algorithmic steps leading to an output from a neural network as follows:

(1) Inputs are x_i, $\forall_i$

(2) Output of hidden layer

$$h_j = \sum_{i=0}^{M} w_{ij} x_i \qquad (8.4a)$$

(3) Assuming a sigmoid is used as the activation function the output of each neuron is

$$z_j = \frac{1}{1 + e^{-h_j}}; \ \forall_j \qquad (8.4b)$$

(4) The output layer provides the following result:

$$b = \sum_{j=0}^{N} \beta_j z_j \qquad (8.4c)$$

(5) The output of the neural network when a sigmoid function is used is

$$y = \frac{1}{1 + e^{-b}} \qquad (8.4d)$$

8.2.1 Activation Functions

Activation functions feature prominently in the analysis of neural networks and deep learning networks. In this section, several activation functions have been discussed comparatively. This allows informed choice of which activation is to be used for applications. An activation function is a nonlinear function used to transform and squash its input data of any range to a new range that lies within zero (0) and 1. For some activation functions, the output range lies between ± 1 and -1. Typical activation functions include the sigmoid whose output range is [0,1], hyperbolic tangent $[-1, 1]$, rectified linear unit (ReLU) [0,x] where x could be a value more than one and leaky ReLU [0,ax], where a is a constant. In terms of ANN, an activation function should be differentiable. That helps to support updating of network parameters through the backpropagation algorithm.

8.2.1.1 Sigmoid

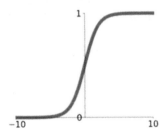

$$f(x) = \frac{1}{1 + e^{-x}} \tag{8.5}$$

The gradient of this function is

$$\frac{\partial f(x)}{\partial x} = f(x) \times (1 - f(x)) \tag{8.6}$$

The first problem with the sigmoid function is that when x is large the gradient flattens. This has the effect of killing off the neural network gradient being passed down from the output into the input during backpropagation exercise. The second problem with the sigmoid function is that the gradient is not centred on zero. When $x = 0$, the gradient is 0.5. Indeed, for any value of x, the gradient is still positive, meaning that it will affect gradients being passed down from the output to the input. The third problem is that it is expensive to compute an exponential function.

8.2.1.2 Hyperbolic tangent

This is a scaled sigmoid function with expression

$$f(x) = \tanh(x) = \frac{2}{1 + e^{-2x}} - 1 = 2\ sigmoid(2x) - 1 \qquad (8.7)$$

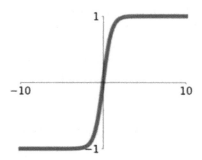

The hyperbolic tangent solves one problem associated with the sigmoid function. It is zero-centred. It however still has with it the problem of its gradient being saturated at high values of x and kills gradients being passed down from the output to the input. The gradient of the function is

$$\frac{\partial f(x)}{\partial x} = 1 - (f(x))^2 \qquad (8.8)$$

8.2.1.3 Rectified Linear Unit (ReLU)

The rectified linear unit or ReLU is differentiable and the gradient is also centred. It is also computationally efficient and converges a lot faster than the sigmoid and the hyperbolic tangent. It is considered more suited to biological signal processing than the sigmoid and hyperbolic tangent. For these reasons, it is more commonly used in ANN applications.

$$f(x) = \max(0, x) \qquad (8.9)$$

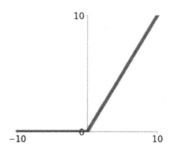

From the graph of the ReLU, the gradient is saturated when x is negative. The gradient is also not centred, thus maintaining some of the problems associated with the other activation functions. For these reasons, ReLU will normally act as if your data is divided into two parts. In one part, the ReLU works well in region so-called as 'active ReLU', and in another region, it does not perform well and called 'dead ReLU' region. This is a region where the ReLU will not activate and not lead to updates of weights. For some data, this problem could be reduced by using small positive biases. For others, doing so would not help at all. This leads to the notion of 'leaky ReLU'.

8.2.1.4 Leaky ReLU
The leaky ReLU modifies the ReLU activation function by initially using a small bias in the ReLU expression as follows:

$$f(x) = \max(10^{-2}x, x) \tag{8.10}$$

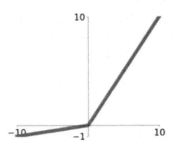

The rectified linear unit now has a slope when its input is negative. This effect can be generalized by using the Parametric Rectifier.

8.2.1.5 Parametric rectifier
The parametric rectifier (PReLU) is a generalization of the ReLU. The activation function is

$$f(x) = \max(ax, x) \tag{8.11}$$

Another generalization of this form of activation function is the exponential linear units described next.

Exponential Linear Units (ELU)

$$f(x) = \begin{cases} x & x \geq 0 \\ -a(1 - \exp(x)) & x \leq 0 \end{cases} \tag{8.12}$$

This function has a negative saturation region.

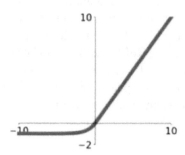

Is there then any activation function which does not saturate? One of such solutions was given by Goodfellow et al. in 2013 and is called the Maxout 'Neuron'.

8.2.1.6 Maxout neuron

The Maxout neuron generalizes the ReLU variants of activation function. Its core strengths are that it does not saturate, has a linear regime and its slope does not die. It is given by the expression

$$f(x) = \max(w_1^T x + b_1, w_2^T x + b_2) \tag{8.13}$$

8.2.1.7 The Gaussian

The Gaussian activation function aims to reduce the problems associated with the activation functions discussed earlier. The gradient of the Gaussian function is centred and has a very small region around the axis of symmetry where the gradient saturates.

$$f(x) = e^{-x^2/2} \tag{8.14a}$$

The gradient of this function is

$$\frac{\partial f(x)}{\partial x} = -xe^{-x^2/2} \tag{8.14b}$$

8.2.1.8 Error calculation

Consider that the sigmoid function is used as the activation function. The sigmoid is a continuous function and differentiable. In this section, we show how to obtain its derivative and then use it.

The derivative of the sigmoid function is obtained using the ratio rule and is

$$\frac{d\sigma(x)}{dx} = \frac{d}{dx}\left(\frac{1}{1+e^{-x}}\right) = \frac{e^{-x}}{(1+e^{-x})^2}$$
$$= \frac{1+e^{-x}}{(1+e^{-x})^2} - \frac{e^{-x}}{(1+e^{-x})^2} \tag{8.15}$$

This is simplified further to be

$$\frac{d\sigma(x)}{dx} = \frac{\sigma^2(x)}{\sigma(x)} - \sigma^2(x) = \sigma(x) - \sigma^2(x)$$
$$= \sigma(x)(1-\sigma(x)) \tag{8.16}$$

This section shows how to obtain the derivative of the error of the neuron output

The output y of the neuron is

$$o = \sigma(wx) \tag{8.17}$$

The derivative of the output of the neuron therefore is

$$\frac{d\sigma(wx)}{dx} = \sigma(wx)(1-\sigma(wx)) \tag{8.18}$$

By including the bias term in the neuron, we obtain the following diagram and output value as

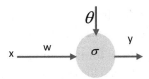

The output of the neuron now needs to include the bias to the neuron

$$o = \sigma(wx + \theta) \tag{8.19}$$

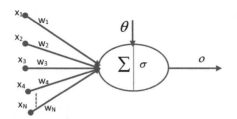

In a typical neuron in a neural network, there could be many inputs to it of the form shown in the figure. The output of the neuron is

$$o = \sigma(w_1 x_1 + w_2 x_2 + w_3 x_3 + \cdots + w_N x_N + \theta) \qquad (8.20)$$

The neural network in our example contains an input layer, a hidden layer and an output layer denoted by I, J and K, respectively, in the following diagram.

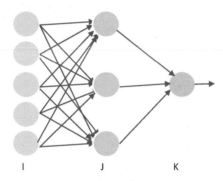

The full notations being used to undertake the backpropagation algorithm include the following. The index l refers to the layer of interest and i, j, and k are indices at the input, hidden and output layers, respectively.

x_j^l is the input to node j of layer l
w_{ij}^l is the weight from node i of layer $l-1$ to node j of layer l
θ_j^l is the bias of node j in layer l
o_j^l is the output of node j in layer l
t_j is the desired value of node j of the output layer

This notation will be used for describing the backpropagation model.

The error equation between the output of an output neuron and the desired value is given by the equation

$$E = \frac{1}{2} \sum_{k=1}^{K} (o_k - t_k)^2 \qquad (8.21)$$

The target values t_j are derived from the training set. The next step in the process is to calculate the rate of change of the error with respect to the weights linking the previous layer to the current layer. Two cases need to be considered. First, we consider when the error derivative is derived at a hidden layer and second when it is derived at the output layer.

8.2.1.9 Output layer node

At an output layer node, the derivative of the error is given by the expression

$$\frac{\partial E}{\partial w_{jk}} = \frac{\partial}{\partial w_{jk}} \left[\frac{1}{2} \sum_{k=1}^{K} (o_k - t_k)^2 \right] \tag{8.22}$$

Using the chain rule with t_k as a constant, this derivative is therefore

$$\frac{\partial E}{\partial w_{jk}} = (o_k - t_k) \times \frac{\partial}{\partial w_{jk}} o_k \tag{8.23}$$

Substituting into o_k, this becomes

$$\frac{\partial E}{\partial w_{jk}} = (o_k - t_k) \times \frac{\partial}{\partial w_{jk}} \sigma(x_k) \tag{8.24}$$

For the next step, it is required to differentiate the activation function. Using, for example, the sigmoid as the activation function, we may now substitute for its derivative. The chain rule also means that we must differentiate the argument of the sigmoid function. The resulting expression is

$$\frac{\partial E}{\partial w_{jk}} = (o_k - t_k) \times \sigma(x_k)(1 - \sigma(x_k)) \frac{\partial}{\partial w_{jk}} x_k$$
$$= (o_k - t_k) \times \sigma(x_k)(1 - \sigma(x_k)) o_j \tag{8.25}$$

Since $\sigma(x_k) = o_k$, we can rewrite the result as

$$\frac{\partial E}{\partial w_{jk}} = (o_k - t_k) \times o_k(1 - o_k) o_j \tag{8.26}$$

Let

$$\delta_k = (o_k - t_k) \times o_k(1 - o_k)$$

Then

$$\frac{\partial E}{\partial w_{jk}} = \delta_k o_j \tag{8.27}$$

8.2.1.10 Hidden layer nodes

The derivatives at the hidden layer nodes follow the same form of analysis as in the output layer nodes. This is

$$\frac{\partial E}{\partial w_{ij}} = \frac{\partial}{\partial w_{ij}} \left[\frac{1}{2} \sum_{k \in K} (o_k - t_k)^2 \right] = \sum_{k \in K} (o_k - t_k) \frac{\partial o_k}{\partial w_{ij}} \qquad (8.28)$$

The output node at this layer leads to

$$\frac{\partial E}{\partial w_{ij}} = \sum_{k \in K} (o_k - t_k) \frac{\partial}{\partial w_{ij}} \sigma(x_k) \qquad (8.29)$$

This derivative is

$$\frac{\partial E}{\partial w_{ij}} = \sum_{k \in K} (o_k - t_k) \times \sigma(x_k)(1 - \sigma(x_k)) \frac{\partial}{\partial w_{ij}} x_k \qquad (8.30)$$

Using the chain rule, we can break this derivative down to two components as a product of partial derivatives as follows

$$\frac{\partial}{\partial w_{ij}} x_k = \frac{\partial x_k}{\partial o_j} \frac{\partial o_j}{\partial w_{ij}} \qquad (8.31)$$

Hence

$$\frac{\partial E}{\partial w_{ij}} = \sum_{k \in K} (o_k - t_k) \times \sigma(x_k)(1 - \sigma(x_k)) \frac{\partial x_k}{\partial o_j} \frac{\partial o_j}{\partial w_{ij}} \qquad (8.32)$$

This is

$$\frac{\partial E}{\partial w_{ij}} = \sum_{k \in K} (o_k - t_k) \times o_k \times (1 - o_k) \times \frac{\partial x_k}{\partial o_j} \frac{\partial o_j}{\partial w_{ij}} \qquad (8.33)$$

To simplify this further, since the partial derivative $\frac{\partial x_k}{\partial o_j} = w_{jk}$ and the partial derivative $\frac{\partial o_j}{\partial w_{ij}}$ are independent of k and therefore do not need to be under the summation sign, the derivative at the hidden node is given by the expression

$$\frac{\partial E}{\partial w_{ij}} = \frac{\partial o_j}{\partial w_{ij}} \sum_{k \in K} (o_k - t_k) \times o_k \times (1 - o_k) \times w_{jk} \qquad (8.34)$$

To further reduce this expression, we need the following equivalents:

$$\delta_k = (o_k - t_k) \times o_k \times (1 - o_k) \tag{8.35}$$

$$\frac{\partial o_j}{\partial w_{ij}} = o_j \times (1 - o_j)o_i \tag{8.36}$$

Therefore

$$\frac{\partial E}{\partial w_{ij}} = o_j \times (1 - o_j) \times o_i \sum_{k \in K} (o_k - t_k) \times o_k \times (1 - o_k) \times w_{jk} \tag{8.37}$$

Therefore

$$\frac{\partial E}{\partial w_{ij}} = o_i \times o_j \times (1 - o_j) \times \sum_{k \in K} \delta_k \times w_{jk} \tag{8.38}$$

Using a similar definition as in the output layer, we rewrite the left-hand side of the above equation as

$$\frac{\partial E}{\partial w_{ij}} = o_i \delta_j \tag{8.39}$$

Therefore

$$\delta_j = o_j \times (1 - o_j) \times \sum_{k \in K} \delta_k \times w_{jk} \tag{8.40}$$

8.2.1.11 Summary of derivations

In summary, four equations define how partial derivatives at the output layer and hidden layers are obtained. They are

For the output layer $(k \in K)$

$$\frac{\partial E}{\partial w_{jk}} = o_j \delta_k$$

$$\delta_k = (o_k - t_k) \times o_k \times (1 - o_k)$$

For the hidden layer $(j \in J)$

$$\frac{\partial E}{\partial w_{ij}} = o_i \delta_j$$

Therefore

$$\delta_j = o_j \times (1 - o_j) \times \sum_{k \in K} \delta_k \times w_{jk}$$

The impact of the bias term on these results needs to be ascertained. For the output node, the derivative of the output to the bias term is always unity. That is

$$\frac{\partial o}{\partial \theta} = 1$$

For all the hidden layers, we find that

$$\frac{\partial E}{\partial \theta} = \delta_l$$

(1) The summary of the backpropagation algorithm is therefore as follows. Run the network to have an output at the output layer. Then

(2) For every output node, calculate

$$\delta_k = (o_k - t_k) \times o_k \times (1 - o_k) \tag{8.41}$$

(3) For every hidden layer node, calculate

$$\delta_j = o_j \times (1 - o_j) \times \sum_{k \in K} \delta_k \times w_{jk} \tag{8.42}$$

(4) Calculate the change in weights using the expressions

$$\left. \begin{array}{l} \nabla w = -\eta \delta_l o_{l-1} \\ \nabla \theta = -\eta \delta_l \end{array} \right\} \tag{8.43}$$

η is called the learning rate.

(5) Update the weights and bias using the expressions

$$\left. \begin{array}{l} w + \nabla w \to w \\ \theta + \nabla \theta \to \theta \end{array} \right\} \tag{8.44}$$

The arrow ($\to$) means replace what is in the right-hand side of the arrow with what is in the left-hand side of the arrow. The negative signs in step (5) indicate that the update is with respect to a steepest descent gradient as opposed to steepest ascent gradients (which should be positive sign).

9

Training of Neural Networks

9.1 Introduction

In Chapter 6, a brief introduction of how to use computation graphs is given. Computation graphs simplify backpropagation algorithms. In this chapter, we derive the expressions for backpropagation and apply it using computation graphs. A step-by-step demonstration of how to train a neural network of three layers (input, hidden and output layers) is given to take the reader through the learning process. It is hoped that a better appreciation in use, understanding of the functions and modification of modern data analytic software will emerge.

9.2 Practical Neural Network

A practical neural network contains at least three layers, an input layer, a hidden layer and an output layer. This is shown in Figure 9.1.

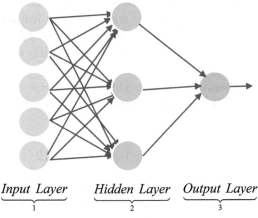

Input Layer   *Hidden Layer*   *Output Layer*
       1   2   3

Figure 9.1   Neural network.

In Figure 9.1, the input layer neurons are labelled x_m. There are M of them. The hidden layer neurons are designated by h_n. There are N of them. The hidden layer is connected to the output layer. Many ANN applications limit the number of layers to three. In deep learning applications, many more layers are involved and lead to deeper learning of underlying processes and better intelligence from the ANN.

9.3 Backpropagation Model

9.3.1 Computational Graph

The backpropagation model is an iterative method used in neural networks for updating the weights in the network. It computes the partial derivative of the output of each perceptron with respect to each of its inputs and continues the process until the overall inputs to the neural network is reached. This section introduces a familiar computational graph (circuit) in many online tutorials on the explanation of back propagation.

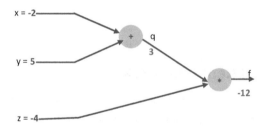

In electronic engineering parlance, the computational graph is actually a circuit connecting one addition gate and one multiplication gate. The inputs to the addition gate are x and y and the third input is z. The z input is connected directly to the multiplication gate. The output of the addition gate is q, and the overall output is f.

The output expression for the addition gate $q = x + y = 3$

The output of expression for the multiplication gate:

$$f = (x + y)z$$
$$f = qz = 3 \times (-4) = -12$$

Backpropagation computes the derivatives of weights and use them to update the value of weights to train neural networks to learn and produce realistic predictions or outputs. We compute the following partial derivatives in the computational graph. Starting from the output of the graph, the following derivatives can be written

$$\frac{\partial f}{\partial f} = 1; \quad \frac{\partial q}{\partial x} = 1; \quad \frac{\partial q}{\partial y} = 1$$

$$\frac{\partial f}{\partial q} = z; \quad \frac{\partial f}{\partial z} = q$$

These partial derivatives are required for computing the partial derivative of the output f with respect to the inputs: $\frac{\partial f}{\partial x}, \frac{\partial f}{\partial y}, \frac{\partial f}{\partial z}$. In the figure, values of the partial derivatives are given as follows:

$$\frac{\partial f}{\partial x} = \frac{\partial f}{\partial q} \cdot \frac{\partial q}{\partial x} = z = -4$$

$$\frac{\partial f}{\partial y} = \frac{\partial f}{\partial q} \cdot \frac{\partial q}{\partial y} = z = -4$$

$$\frac{\partial f}{\partial q} = 3$$

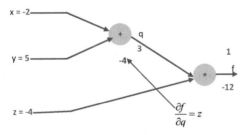

We walk our way back to compute the partial derivative of the output with respect to the inputs to the network. They are:

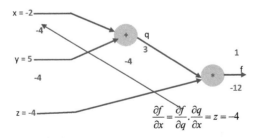

Also for the partial derivative of f with respect to y, we have:

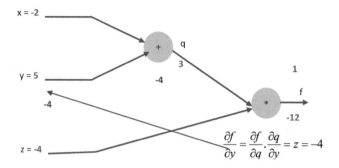

$$\frac{\partial f}{\partial y} = \frac{\partial f}{\partial q} \cdot \frac{\partial q}{\partial y} = z = -4$$

The partial derivative with respect to z is given in the following diagram:

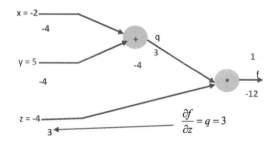

$$\frac{\partial f}{\partial z} = q = 3$$

In a nutshell, the back propagation model repeatedly uses the differentiation chain rule and progresses from the output back to the input. This propagates changes at the output to the input so that weights could be changed if necessary. The derivatives in these diagrams are with respect to their immediate neighboring variables only. This is further illustrated as

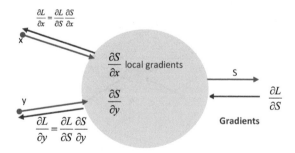

Repeated use of the chain rule leads to the partial derivatives at the inputs to the network leading to propagation of the errors made at the output into the

inputs to enable variation of the weights. Note that at the output, if there is no further stage or node, the derivative is $\frac{\partial S}{\partial S} = 1$.

9.4 Backpropagation Example with Computational Graphs

In this section, we consider another example. In this case, activation function is connected to the neuron to produce a decision value. Consider the following computation graph depicting the forward path of a neural network. It consists of inputs with weights and activation function (a sigmoid) at the output. The calculation is shown from input to output.

For the forward path:

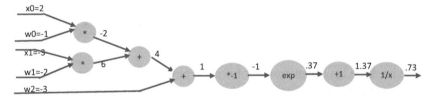

Consider a neural network as shown above that uses the sigmoid activation function. The inputs to and output from the network are therefore related by the activation function

$$f(x, w) = \frac{1}{1 + e^{-(w_0 x_0 + w_1 x_1 + w_2)}}$$

Following the network behaviour from the input to the output as defined in the Figure, the network will output a value of 0.73, a positive value and hence the neuron will fire or be triggered.

The following derivatives will be used for back propagation from the output in relation to the activation function.

Function f(x)	Derivative
$f(x) = e^x$	$\dfrac{\partial f(x)}{\partial x} = e^x$
$f_a(x) = ax$	$\dfrac{\partial f_a(x)}{\partial x} = a$
$f(x) = \dfrac{1}{x}$	$\dfrac{\partial f(x)}{\partial x} = -\dfrac{1}{x^2}$
$f_c(x) = c + x$	$\dfrac{\partial f_c(x)}{\partial x} = 1$

9.5 Back Propagation

The objective of backpropagation is to estimate how to update the weights in a neural network by using partial derivatives. Backpropagation starts from the output of the network and walks back to the input. At the output

$$\frac{\partial f}{\partial f} = 1$$

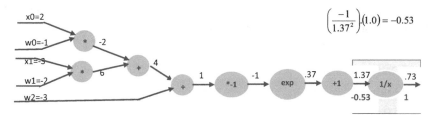

We have the upstream gradient from the output coming down is the red value of 1, and we also know that the derivative of the inverse of x is:

$$f(x) = \frac{1}{x} \quad \frac{\partial f(x)}{\partial x} = -\frac{1}{x^2}$$

Therefore, the product of partial derivative coming down is:

$$\left(\frac{-1}{x^2}\right) \cdot (1.0) = \left(\frac{-1}{1.37^2}\right) \cdot (1.0) = -0.53$$

For the next step in the backpropagation downwards, the node is a $+1$ node and the derivative is that of:

$$f_c(x) = c + x \quad \frac{\partial f_c(x)}{\partial x} = 1$$

Therefore, it is the product of the local gradient and the upstream gradient: $\cdot(1.0)(-0.53) = -0.53$

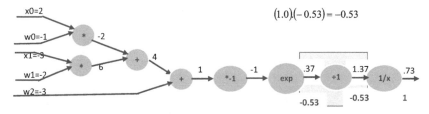

The next step of the backpropagation involves the derivatives

$$f(x) = e^x \quad \frac{\partial f(x)}{\partial x} = e^x$$

This is given by the product of the upstream gradient and the local gradient which with x $= -1$ is

$$(-0.53)(e^{-x}) = -0.53(e^{-1}) = -0.2$$

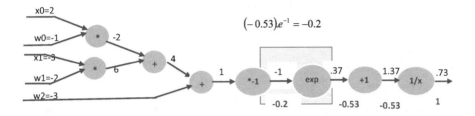

The next step of the backpropagation requires the product of the upstream gradient and the local gradient, which is determined from the expression of a product node and is:

$$f_c(x) = ax \quad \frac{\partial f_c(x)}{\partial x} = a$$

The value of a $= -1$ and the product of gradients is: $(-1.0) \cdot (-0.2) = 0.2$

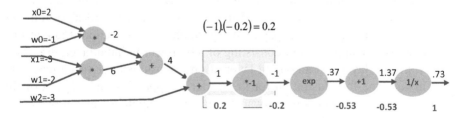

For the two branches linking the sum node, the backpropagation is computed as the product of the local gradient (+1) and the upstream gradient, which is 0.2. It happens in this case that both products are:

$$(1) \cdot (0.2) = 0.2$$

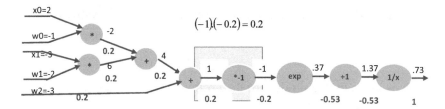

The remaining gradients at the input branches are computed using the computational graph rule of a product of variables and the upstream gradients for alternate branch values as follows:

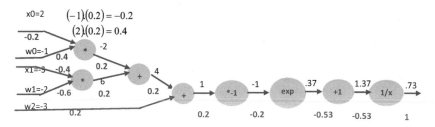

Obviously, the use of computational graph simplifies computation of the backpropagation gradients because it involves only plugging in the values of the local and upstream gradients and forming a product.

9.6 Practical Training of Neural Networks

Training of neural network is the process by which weights attached to the synapses or links are adapted first by using the forward algorithm and followed with the backpropagation algorithm. In this section, training of a three-layer neural network consisting of a two-neuron input layer, three neuron hidden layer and an output layer of one neuron is undertaken manually. This is done for illustration for the reader to appreciate the steps normally taken by a neural network software. A manual computation also provides the basis for being able to understand when things go wrong during training as an aid to resolving such problems.

During the forward algorithm, a set of weights are selected either randomly or using some criteria. The weights are applied to the input data forward with calculation of output. In the first iteration of the NN, the set of weights may be selected randomly. This is acceptable because during the backpropagation, the measure of error made in the choice of the weights will

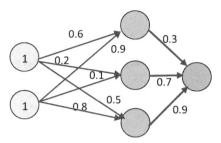

Figure 9.2 Three-layer neural network.

be computed, which will allow adaptation of the weights for the next round of the algorithm.

The objective in the computation of the error is to keep on decreasing the error made in computing the weights. Once the weights become 'stable', then system parameters have been determined to the best approximation.

9.6.1 Forward Propagation

Consider, for example, a simple neural network consisting of two input neurons, three neurons in the hidden layer and one output neuron. This network is shown in Figure 9.2. In general, we will represent our input as a matrix of numbers and the weights as also a matrix. An example of such matrices is given in the following expression.

$$\begin{bmatrix} x_{11} & x_{12} \\ x_{21} & x_{22} \\ x_{31} & x_{32} \\ x_{11} & x_{42} \end{bmatrix} \times \begin{bmatrix} w_{11} & w_{12} & w_{13} \\ w_{21} & w_{22} & w_{23} \end{bmatrix} = Hidden\ Layer = f_h$$

This example is the case for a 4×2 input matrix and a 2×3 weight matrix. The product results in a 4×3 output from the hidden layer neurons. Consider our example in Figure 9.2. In the neural network below, the target is 0.7 at the output neuron. We present the forward propagation in steps.

Step 1: The products of input and weights at the hidden layer

$$X \times W = f_h$$
$$1 \times 0.6 + 1 \times 0.9 = 1.5$$
$$1 \times 0.2 + 1 \times 0.1 = 0.3$$
$$1 \times 0.5 + 1 \times 0.8 = 1.3$$

These results are shown in the hidden layer in red.

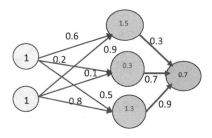

Step 2: Apply Activation Function at Hidden layer

Next, we apply an activation function to the intermediate results in the hidden layer neurons. The activation function is used to transform these results to an output from each neuron. As an example, the sigmoid function is used in this chapter. The outputs of the three neurons in the hidden layer are therefore:

$$h_1 = \frac{1}{1 + e^{-1.5}} = 0.82$$

$$h_2 = \frac{1}{1 + e^{-0.3}} = 0.57$$

$$h_3 = \frac{1}{1 + e^{-1.3}} = 0.79$$

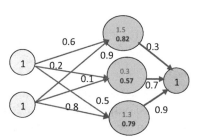

Step 3: Sum outputs from the hidden layer

These results from the activation function process are shown inside the hidden neurons in blue text. The outputs from the hidden layer are given by the three hidden layer neurons and are:

$$h_1 \times v_1 + h_2 \times v_2 + h_3 \times v_3 = f_o$$
$$0.82 \times 0.3 + 0.57 \times 0.7 + 0.79 \times 0.9 = 1.356$$

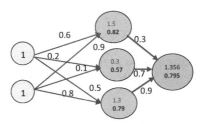

Step 4: Apply Activation Function at Output Neuron

Next we apply the activation function to get an output.

This is: $\frac{1}{1+e^{-f}} = \frac{1}{1+e^{-1.356}} = 0.795 = Output$

From the results so far, we have not achieved the intended target, which was set as 0.7. The obtained result so far is 0.795, which is not too far from the target. To get the result close to the intended target, we now use the backpropagation algorithm to adapt the weights and the repeat the process.

9.6.2 Backward Propagation

Note that the output of the last neuron was obtained by applying the sigmoid function to the sum (f) of the products of weights and inputs. This is

$$Sigmoid\ (sum) = \sigma(f) = \frac{1}{1+e^{-f}} = Output$$

By observing this equation, notice that any change in the sum will cause a change in the output. Also any change in the output will be as a result of the change in the sum. The sum will change if the weights of the links to the neuron change. We will present the back propagation algorithm in steps.

Step 5: Compute Difference Between Target Value and Result

The change in output is given by the difference between the desired value and the actual obtained output. This is for the example in hand:
error $=$ target $-$ result $= 0.7 - 0.795 = -0.095$. We will define this as $\Delta(Output)$ or error.

Step 6: Compute change in sum as Result of change in output

How do we compute the actual change in the sum? The change in the sum is given by the derivative of sigmoid as a function of the sum. This is given by the expression:

$$\sigma'(f) = \frac{d(f)}{d(Output)} = \frac{\Delta(f)}{error}$$

This means that

$$\sigma'(f) \times error = \Delta(f)$$

We will use this equation often in the analysis. Note that σ' is the derivative of the activation function, which in our case is the Sigmoid.

$$\frac{d\sigma(x)}{dx} = \sigma(x)(1 - \sigma(x))$$

The first output from the backpropagation is therefore

$$\Delta f = \left[\frac{1}{1 + e^{-x}} \left(1 - \frac{1}{1 + e^{-x}} \right) \right] \times error$$

$$= \left[\frac{1}{1 + e^{-1.356}} \left(1 - \frac{1}{1 + e^{-1.356}} \right) \right] \times (-0.095) = -0.019$$

9.6.2.1 Adapting the weights
Step 7: *Adapt the Weights (between hidden layer and output layer)*

In the network, the output layer contains only one neuron. In this section, we use $\Delta f = -0.019$ in the derivative of the output sum expression. This is to allow us to obtain the new weights (taking into of account changed weights). Since the output function is a product the weights between the hidden layer and output with the hidden layer results, we can adapt the weights. The output sum f_o is

$$h_1 \times v_1 + h_2 \times v_2 + h_3 \times v_3 = f_o$$
$$H \times v_k = f_o$$

At the hidden layer

$$H = \frac{\Delta f_o}{\Delta v_k}; \qquad \begin{matrix} h_1 = 0.82 \\ h_2 = 0.57 \\ h_3 = 0.79 \end{matrix}$$

Therefore, the changes in weights between hidden and output layer are:

$$\Delta v_1 = \frac{\Delta f_o}{h_1} = \frac{-0.019}{0.82} = -0.02317$$

$$\Delta v_2 = \frac{\Delta f_o}{h_2} = \frac{-0.019}{0.57} = -0.03333$$

$$\Delta v_3 = \frac{\Delta f_o}{h_3} = \frac{-0.019}{0.79} = -0.02405$$

The new weights are therefore obtained as follows:

$$v_{knew} = v_{kold} + \Delta v_k$$
$$v_{1new} = v_{1old} + \Delta v_1 = 0.3 - 0.02317 = 0.27683$$
$$v_{2new} = v_{2old} + \Delta v_2 = 0.7 - 0.03333 = 0.6667$$
$$v_{3new} = v_{3old} + \Delta v_3 = 0.9 - 0.02405 = 0.87595$$

Step 8: *Compute the Change in Hidden Layer Sum*

Next, focus on the remaining part of the backpropagation from the hidden layer to the input layer. It can be shown that the change in the hidden layer sum is due to changes in the output sum and the output weights between hidden layer and output layer. It is given by the derivatives of two terms, the derivatives of activation function and the output sum that involves the derivative of the weights. This is given formally with the equation:

$$d(f_h) = \frac{df_o}{v_k} \times \sigma'(f_h)$$
$$df_o = -0.019; \quad v = \begin{bmatrix} 0.3, & 0.7, & 0.9 \end{bmatrix}; \quad f_h = \begin{bmatrix} 1.5, & 0.3, & 1.3 \end{bmatrix}$$

Thus, we can now write in a cryptic form the change in output sum to be

$$d(f_h) = \frac{-0.019}{\begin{bmatrix} 1.5, & 0.3, & 1.3 \end{bmatrix}} \times \sigma'\left(\begin{bmatrix} 0.3, & 0.7, & 0.9 \end{bmatrix}\right)$$

Take each element of this expression, one at a time compute the changes as

$$d(f_h)_1 = \frac{-0.019}{1.5} \times \sigma'(0.3) = -0.00094$$
$$d(f_h)_2 = \frac{-0.019}{0.3} \times \sigma'(0.7) = -0.014$$
$$d(f_h)_3 = \frac{-0.019}{1.3} \times \sigma'(0.9) = -0.0030$$

Therefore, the change in hidden layer sum is

$$d(f_h) = [-0.00094, \quad -0.0014, \quad -0.003]$$

This provides the change in the hidden summations inside the neurons in the hidden layer.

Step 9: *Adapt Weights Between Input and Hidden Layer*

Next, update the weights between the input and the hidden layer. The change in input weights is given by the expression

$$\Delta w_k = \frac{d(f_h)_k}{Input\ data}$$

The input data was [1, 1]. Therefore, the change in weights between the input and hidden layers is given formally as:

$$\Delta w_k = \frac{\begin{bmatrix} -0.00094, & -0.0014, & -0.003 \end{bmatrix}}{\begin{bmatrix} 1, & 1 \end{bmatrix}}; \quad k = 1, 2, 3$$

The result to the above expression is easily obtained by dividing each element by the input which is

$$\Delta w = \begin{bmatrix} -0.00094, & -0.0014, & -0.003, & -0.00094, & -0.0014, & -0.003 \end{bmatrix}$$

The new weights become the old weights plus the corresponding change in weights. The weights are

$$w_{k,new} = w_{k,old} + \Delta w_k$$

Weight Index	Old Value of Weight	Delta Weight	New Weight
W_{11}	0.6	−0.00094	0.599
W_{12}	0.2	−0.0014	0.199
W_{13}	0.5	−0.003	0.497
W_{21}	0.9	−0.00094	0.899
W_{22}	0.1	−0.0014	0.099
W_{23}	0.8	−0.003	0.797

The new weights are smaller in this case than the original ones we chose at the beginning. This concludes the first complete training iteration. These steps of going through the forward propagation followed by backpropagation are normally done several thousand times in practical neural network applications until the output from the network is as close as possible or equal to the desired target output.

9.7 Initialisation of Weights Methods

In our previous example in Section 9.6, weight initialisation was random. A set of non-zero weights were used. These weights may as well be all zeros. Using zero weights will not detract from good training since weights are adapted many times during the training iterations. Weights may also be big or small in values, be a mix of positive and negative values. There are disadvantages in using very small weights. Small weights could cause gradients to be zeros and activation functions becoming zeros. Small weights often lead to limitations in learning by the neural network. Very large weights could also cause activation functions to saturate and gradients become zeros. This also results to no learning by the neural network.

Several initialisation algorithms exist including the Xavier initialisation and batch normalisation to name a few. These two algorithms are described in this section. Understandably, the two of them are recent and used for deep learning network but just as well are useful in the choice of parameters for smaller networks as described in this chapter.

9.7.1 Xavier Initialisation

Small weights cause the variance of the input signal to diminish through the layers as the algorithms go through each layer of the network. When the weights are too large, the variance of the input data increases rapidly at each passing layer. Eventually, the variance becomes too large, tending to infinity and becoming useless. These changes are caused by activation functions such as the sigmoid as they flatten and saturate. Making sure that the network is initialised with the right weights in the right range is thus important but not easily achievable. Xavier initialisation was introduced to help with the right selection of weights.

Often the input data is unprocessed meaning that near to no information about the input is known and hence not sure how to assign the weights. In Xavier [1–3] initialisation, weights are selected from a Gaussian distribution of zero mean and finite variance. This gives us a consistent approach for assigning weights instead of a random approach. The objective of the initialisation is to make sure that at each passing layer of the network, the variance of the input remains the same.

Consider a linear neuron as an example. The linear neuron may be represented with the expression

$$y = w_1 x_1 + w_2 x_2 + \cdots + w_N x_N + b$$

where b is the bias at the neuron. For the variance of this expression to remain the same at each passing layer, the variance of y must be equal to the variance of the right-hand side of the equation. Hence, we write

$$\text{var}(y) = \text{var}(w_1 x_1 + w_2 x_2 + \cdots + w_N x_N + b)$$

Note that the bias is a constant and therefore does not vary. Its variance is zero. The variance of the rest of the equation at the right-hand side is given as follows. Since

$$\text{var}(w_i x_i) = E(x_i)^2 \, \text{var}(w_i) + E(w_i)^2 \, \text{var}(x_i) + \text{var}(w_i) \, \text{var}(x_i)$$

The expectation $E(x)$ is the mean value of the variable x. Since we assumed that both the weights and input values are zero mean Gaussian distributions, the expectations in the above equation vanish. Hence, we have

$$\text{var}(w_i x_i) = \text{var}(w_i) \, \text{var}(x_i)$$

Therefore, we can now write an expression for the $\text{var}(y)$ as products of variances of the inputs and weights as

$$\text{var}(y) = \text{var}(w_1) \, \text{var}(x_1) + \text{var}(w_2) \, \text{var}(x_2) + \cdots + \text{var}(w_N) \, \text{var}(x_N)$$

The weights and inputs are all identically distributed. This means that we can simplify the above equation to the simple equation:

$$\text{var}(y) = N \times \text{var}(w_i) \, \text{var}(x_i)$$

For the variance of y to be equal to the variance of x, the term $N \times \text{var}(w_i)$ is set to unity. Or

$$N \times \text{var}(w_i) = 1$$

This essentially means that

$$\text{var}(w_i) = \frac{1}{N}$$

Every weight used therefore has the same variance as the rest and it is known (1/N) for all the N input neurons. Thus, the weights chosen from a Gaussian distribution have zero mean and variance 1/N individually. Although, more computationally complex Xavier's method originally used an average N consisting of the sum of input and output neurons as:

$$N_{avg} = \frac{N_{input} + N_{output}}{2}$$

and

$$\text{var}(w_i) = \frac{1}{N_{avg}}$$

Using this form of variance also preserves the input and backpropagated signals.

9.7.2 Batch Normalisation

Batch normalisation [4] when used is an additional layer in the neural network used to force input data to have zero mean and unit variance. This is achieved by initially subtracting the mean of the inputs from input data and dividing them by the variance of the process. Batch initialisation is therefore used to force weights to have Gaussian distribution. The process takes the input data to a layer and transforms it into a Gaussian input set.

Consider N large data set of dimension D (N × D), the batch normalisation for the data set is given as follows:

(1) Compute mean of each array (x: N × D)

$$\mu_j = \frac{1}{N} \sum_{i=1}^{N} x_{i,j}$$

(2) Compute the variance of each array

$$\sigma_j^2 = \frac{1}{N} \sum_{i=1}^{N} (x_{i,j} - \mu_j)^2$$

(3) Normalise the input data

$$\hat{x}_{i,j} = \frac{x_{i,j} - \sigma_j}{\sqrt{\sigma_j^2 + \varepsilon}}$$

The variables have dimensions: $\mu, \sigma : D$ and $\hat{x} : N \times D$

(4) Output

$$y_{i,j} = \gamma_j \hat{x}_{i,j} + \beta_j$$

9.8 Conclusion

In conclusion, it is essential to choose weights wisely using either of Xavier or random selection methods. For large data sets, batch normalisation could be applied to input data. In other words, process the data. During the training phase, check that the loss in results is reasonable and that the network is learning. Since the training process involves the use of steepest descent, check that the network is not stuck at a saddle point. For good learning, the loss gradient reduces progressively. Very high learning rate can be inferred from watching the loss gradient value as a function of the epoch. Low learning rates cause very slow reduction of the loss gradient.

References

[1] Xavier Glorot and Yoshua Benglo, "Understanding the difficulty of training deep feedforward neural networks", Proceedings of the Thirteenth International Conference on Artificial Intelligence and Statistics, PMLR, 2010, pp. 249–256.

[2] Kaiming He, Xiangyu Zhang, Shaoqing Ren and Jian Sun, "Delving Deep into Rectifiers: Surpassing Human-Level Performance on ImageNet Classification", arXiv:1502.01852 [cs], Feb. 2015.

[3] Kaiming He, Xiangyu Zhang, Shaoqing Ren and Jian Sun, "Deep Residual Learning for Image Recognition". ArXive-prints, December 2015. URL http://arxiv.org/abs/1512/03385.

[4] Sergey Ioffe and Christian Szegedy, "Batch Normalization: Accelerating Deep Network Training by Reducing Internal Covariate Shift," arXiv:1502.03167 [cs], Feb. 2015.

10

Recurrent Neural Networks

10.1 Introduction

Many commercial data require keeping information of a process from the past, which makes it inefficient to process them using feed forward neural networks. Such data sources include information on stock, text processing and speaker or voice recognition. Therefore, a new version of neural networks that take feedback input is necessary. By keeping memory of previous outputs and making them available to the inputs of a process, new intelligence and decision capabilities are introduced into them. Recurrent (feedback) neural networks provide such algorithm.

Although backpropagation makes it possible to update the weights in a neural network, the algorithm does not provide for feedback and does not facilitate control of the neural network. Also, such networks do not retain any memories of system states. Memory of previous states and events are essential for system control and smarter decision making. Recurrent neural networks are designed to provide feedback from the output to the input of neural networks.

10.2 Introduction to Recurrent Neural Networks

In the context of recurrent neural networks, consider an innovative restaurant with a brilliant idea on how to distinguish itself from competitors by what it cooks for customers depending on the day. The meal of the day is decided by the weather of that day. MamaPut decided that on a sunny day she will cook as its main dish of the day jollof rice and on a rainy day MamaPut will cook pepper soup. In other words, customers should expect to find a feel happy food on both types of days. These two options are represented in Figure 10.1.

The simple neural network has one input and one output, which depends on the weather (Figure 10.2). The weather is the input and the output is food.

Figure 10.1   Meal of the day.

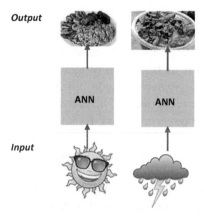

Figure 10.2   Input and output of the neural network.

The two situations represented by Figure 10.2 can be expressed in a way that neural network concepts can be applied. They are expressed as vectors. Let the following vectors represent the different inputs and outputs:

Sunny vector $\begin{bmatrix} 1 \\ 0 \end{bmatrix}$;

Rainy vector $\begin{bmatrix} 0 \\ 1 \end{bmatrix}$;

Jollof rice $\begin{bmatrix} 1 \\ 0 \\ 0 \end{bmatrix}$

Pepper soup $\begin{bmatrix} 0 \\ 1 \\ 0 \end{bmatrix}$

Represent the food or output of the neural network by the observation matrix O and the weather or input of the neural network by the input matrix W, where

$$O = \begin{bmatrix} 1 & 0 \\ 0 & 1 \\ 0 & 0 \end{bmatrix}$$

$$W = \begin{bmatrix} 1 & 0 \\ 0 & 1 \end{bmatrix}$$

Observe that the input weight matrix is an identity matrix. The action or intelligence of the neural network is represented by a matrix multiplication.

$$O \times W = \begin{bmatrix} 1 & 0 \\ 0 & 1 \\ 0 & 0 \end{bmatrix} \times \begin{bmatrix} 1 & 0 \\ 0 & 1 \end{bmatrix} = \begin{bmatrix} 1 & 0 \\ 0 & 1 \\ 0 & 0 \end{bmatrix}$$

This neural network is shown for each of the two input weight vectors. For the sunny day, the network is:

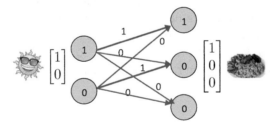

For the rainy day, the neural network is:

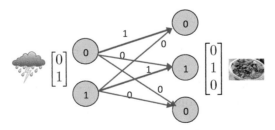

Since the input weight matrix is an identity matrix, the output (food matrix) is maintained when the input weights are multiplied and passed through the artificial neural network. Indeed, by multiplying the output matrix by a vector of the weather (weight W_j; $j \in (1,2)$) matrix, the results are always the appropriate food (vector) as in these examples:

$$O \times W_1 = \begin{bmatrix} 1 & 0 \\ 0 & 1 \\ 0 & 0 \end{bmatrix} \times \begin{bmatrix} 1 \\ 0 \end{bmatrix} = \begin{bmatrix} 1 \\ 0 \\ 0 \end{bmatrix} = \text{(image)}$$

and

$$O \times W_2 = \begin{bmatrix} 1 & 0 \\ 0 & 1 \\ 0 & 0 \end{bmatrix} \times \begin{bmatrix} 0 \\ 1 \end{bmatrix} = \begin{bmatrix} 0 \\ 1 \\ 0 \end{bmatrix} = \text{(image)}$$

10.3 Recurrent Neural Network

So far, the network has no memory and cannot remember what food was cooked yesterday if it was a rainy or sunny day. Memory can be built into the neural network using feedback. The output is fed back into the input to inform the neural network what output was obtained on previous days. When that is done, the network can infer from the output food of the previous days and produce the appropriate food for the day at hand. This means that the network does not anymore need the weather matrix as input, it rather takes input from output feedback. To illustrate this properly, a third input food is added, so that MamaPut provides a menu consisting of jollof rice, pepper soup and suya the Nigerian roast spiced meat.

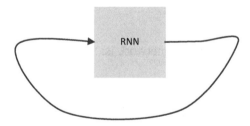

The recurrent neural network produces a regular output of jollof rice, pepper soup and suya as follows:

In the sequence, the RNN, for example, produces jollof rice output, which in the next round becomes input feedback and the output is pepper soup. Next the pepper soup is fed back and the output becomes suya. The next time the output becomes jollof rice again and the sequence repeats. To achieve this, we need a third vector for suya and is given by the vector

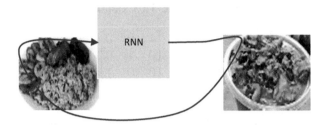

When pepper soup is output, jollof rice is input feedback. When suya is output, pepper soup is input feedback.

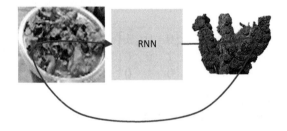

The third time, the input is suya and the output is jollof rice.

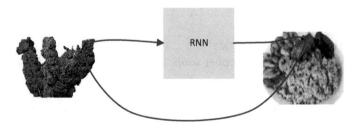

To represent the neural network mathematically, we need a third vector for the suya input and output. The input vector for suya is

$$\begin{bmatrix} 0 \\ 0 \\ 1 \end{bmatrix}$$

The output matrix for the RNN is

$$O = \begin{bmatrix} 0 & 0 & 1 \\ 1 & 0 & 0 \\ 0 & 1 & 0 \end{bmatrix}$$

The input matrix for the RNN is the 3×3 identity matrix

$$W = \begin{bmatrix} 1 & 0 & 0 \\ 0 & 1 & 0 \\ 0 & 0 & 1 \end{bmatrix}$$

Example:

(a) What is the output of the above RNN when the input is jollof rice?
(b) What is the output of the RNN when the input is suya?
(c) What should the output be when the input is pepper soup?

Solution:

(a)

$$O \times W_1 = \begin{bmatrix} 0 & 0 & 1 \\ 1 & 0 & 0 \\ 0 & 1 & 0 \end{bmatrix} \times \begin{bmatrix} 1 \\ 0 \\ 0 \end{bmatrix} = \begin{bmatrix} 0 \\ 1 \\ 0 \end{bmatrix}$$

The output is pepper soup.

(b) When the input is suya, the output expression is

$$O \times W_3 = \begin{bmatrix} 0 & 0 & 1 \\ 1 & 0 & 0 \\ 0 & 1 & 0 \end{bmatrix} \times \begin{bmatrix} 0 \\ 0 \\ 1 \end{bmatrix} = \begin{bmatrix} 1 \\ 0 \\ 0 \end{bmatrix}$$

The output is jollof rice!

(c) When the input is pepper soup, the output is given by the expression

$$O \times W_2 = \begin{bmatrix} 0 & 0 & 1 \\ 1 & 0 & 0 \\ 0 & 1 & 0 \end{bmatrix} \times \begin{bmatrix} 0 \\ 1 \\ 0 \end{bmatrix} = \begin{bmatrix} 0 \\ 0 \\ 1 \end{bmatrix}$$

The output is suya. These results show that this RNN is a linear mapping between its inputs and outputs. It maps pepper soup to suya, suya to jollof rice and jollof rice to pepper soup.

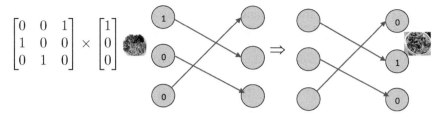

$$\begin{bmatrix} 0 & 0 & 1 \\ 1 & 0 & 0 \\ 0 & 1 & 0 \end{bmatrix} \times \begin{bmatrix} 1 \\ 0 \\ 0 \end{bmatrix}$$

The recurrent neural network really looks like the next figure with outputs connected to the inputs directly.

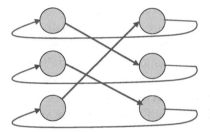

The feedbacks serve to provide memory for the network and hence a bit of more intelligence for it. We can build more intelligence into the RNN by adding the weather process, which will be used as further cue for decision making. This cue for MamaPut is that on a sunny day she re-offers the meal the day before and if it is rainy a feel good dish, pepper soup is cooked. If the next day it is rainy, again she cooks another feel good dish, suya.

This form of neural networks provides two sources of intelligence for decision making, an internal intelligence and external intelligence. The internal intelligence in the form of the food matrix and the external intelligence is in the form of the weather matrix. As will be shown shortly, there is a third source of intelligence, the feedback from output to input and this will be shown as a rotation of the food matrix.

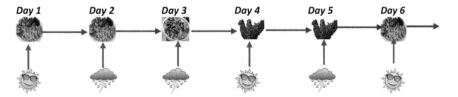

Two matrices, the food and weather matrices, are therefore involved in the processing of this RNN. The food matrix is

$$
\begin{bmatrix} O \\ rot(o) \end{bmatrix} + \begin{bmatrix} W \\ rot(W) \end{bmatrix} = \begin{bmatrix} 1 & 0 & 0 \\ 0 & 1 & 0 \\ 0 & 0 & 1 \\ 0 & 0 & 1 \\ 1 & 0 & 0 \\ 0 & 1 & 0 \end{bmatrix} + \begin{bmatrix} 1 & 0 \\ 1 & 0 \\ 1 & 0 \\ 0 & 1 \\ 0 & 1 \\ 0 & 1 \end{bmatrix}
$$

Notice that both the observation matrix O and the input weight matrix W are divided into two halves. The second half is a one-time-step left-handed rotation of each row of the observation and weight (weather) matrices. What happens when the observation (food) matrix is multiplied by a food vector? An example is given in this section. Let the observation matrix be multiplied by a food vector and in this example the jollof rice vector. The result is a 6×1 vector in which the first half is the jollof rice vector and the second half is the pepper soup vector.

$$
\begin{bmatrix} 1 & 0 & 0 \\ 0 & 1 & 0 \\ 0 & 0 & 1 \\ 0 & 0 & 1 \\ 1 & 0 & 0 \\ 0 & 1 & 0 \end{bmatrix} \times \begin{bmatrix} 1 \\ 0 \\ 0 \end{bmatrix} = \begin{bmatrix} 1 \\ 0 \\ 0 \\ 0 \\ 1 \\ 0 \end{bmatrix}
$$

The first half of the result vector is the vector for the food for today, and the second half is the vector for the food the next day. By repeating this process two times more we notice that the RNN is predicting the food for the next day correctly. That is

$$
\begin{bmatrix} 1 & 0 & 0 \\ 0 & 1 & 0 \\ 0 & 0 & 1 \\ 0 & 0 & 1 \\ 1 & 0 & 0 \\ 0 & 1 & 0 \end{bmatrix} \times \begin{bmatrix} 0 \\ 1 \\ 0 \end{bmatrix} = \begin{bmatrix} 0 \\ 1 \\ 0 \\ 0 \\ 0 \\ 1 \end{bmatrix}
$$

Lastly, the third multiplication gives

$$
\begin{bmatrix} 1 & 0 & 0 \\ 0 & 1 & 0 \\ 0 & 0 & 1 \\ 0 & 0 & 1 \\ 1 & 0 & 0 \\ 0 & 1 & 0 \end{bmatrix} \times \begin{bmatrix} 0 \\ 0 \\ 1 \end{bmatrix} = \begin{bmatrix} 0 \\ 0 \\ 1 \\ 1 \\ 0 \\ 0 \end{bmatrix}
$$

Clearly, the RNN has again predicted the food for today and the next day correctly. We can provide a more conclusive result another way. This could be done by using the weather matrix and vector to help decide what food should be cooked today or tomorrow. This is illustrated next by multiplying the weather matrix with the sunny day vector, which is:

$$
\begin{bmatrix} 1 & 0 \\ 1 & 0 \\ 1 & 0 \\ 0 & 1 \\ 0 & 1 \\ 0 & 1 \end{bmatrix} \times \begin{bmatrix} 1 \\ 0 \end{bmatrix} = \begin{bmatrix} 1 \\ 1 \\ 1 \\ 0 \\ 0 \\ 0 \end{bmatrix}
$$

Same day

Next day

We use the next weather vector that for rainy day to check if the matrix multiplication again provides a good guide of what food to cook. The result is

$$
\begin{bmatrix} 1 & 0 \\ 1 & 0 \\ 1 & 0 \\ 0 & 1 \\ 0 & 1 \\ 0 & 1 \end{bmatrix} \times \begin{bmatrix} 0 \\ 1 \end{bmatrix} = \begin{bmatrix} 0 \\ 0 \\ 0 \\ 1 \\ 1 \\ 1 \end{bmatrix}
$$

Same day

Next day

So far, we have not used any activation functions in the analysis. Activation function will become necessary if the food and weather matrices results are combined as follows:

$$
\begin{bmatrix} 1 \\ 0 \\ 0 \\ 0 \\ 1 \\ 0 \end{bmatrix} \text{(Same day / Next day)} \quad + \quad \begin{bmatrix} 0 \\ 0 \\ 0 \\ 1 \\ 1 \\ 1 \end{bmatrix} \text{(Same day / Next day)} \quad = \quad \begin{bmatrix} 1 \\ 0 \\ 0 \\ 1 \\ 2 \\ 1 \end{bmatrix}
$$

Now if we use an activation function like a sigmoid or ReLU to choose the maximum value as output, it becomes very clear that the food for the next day should be cooked, meaning cook pepper soup, which matches the feel warm food for a rainy day. The activation function basically selects the largest and sets it to $+1$ and the rest to 0. This is

$$
\begin{bmatrix} 0 \\ 0 \\ 0 \\ 0 \\ 1 \\ 0 \end{bmatrix}.
$$

By merging the two halves of this vector, we obtain the more realistic vector

$$
\begin{bmatrix} 0+0 \\ 0+1 \\ 0+0 \end{bmatrix} = \begin{bmatrix} 0 \\ 1 \\ 0 \end{bmatrix},
$$

meaning that the food for the next day (pepper soup) should be cooked.

This result may be demonstrated another way by directly using the 6×1 vector with an activation vector which is a concatenation of two identity matrices as

$$
[I \mid I] \times \begin{bmatrix} 0 \\ 0 \\ 0 \\ 0 \\ 1 \\ 0 \end{bmatrix} = \begin{bmatrix} 1 & 0 & 0 & 1 & 0 & 0 \\ 0 & 1 & 0 & 0 & 1 & 0 \\ 0 & 0 & 1 & 0 & 0 & 1 \end{bmatrix} \times \begin{bmatrix} 0 \\ 0 \\ 0 \\ 0 \\ 1 \\ 0 \end{bmatrix} = \begin{bmatrix} 0 \\ 1 \\ 0 \end{bmatrix} =
$$

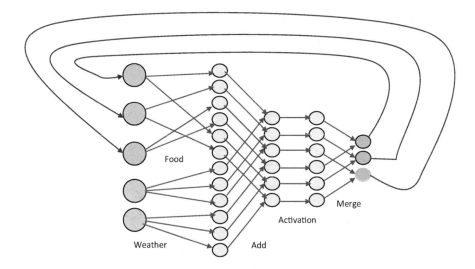

The full RNN is shown consisting of the Food and Weather sections. Behaviours of these two sections are summed followed with an activation process before results are merged. Feedbacks are from the output food nodes to the input food nodes. The operation of this RNN is shown in the next sets of figures with food and weather inputs. The following inputs were applied to the network:

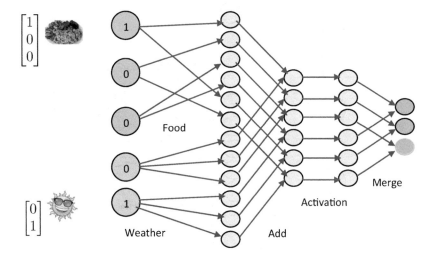

We step through the network to show the processing flow. During the second step, the following information is available in the network.

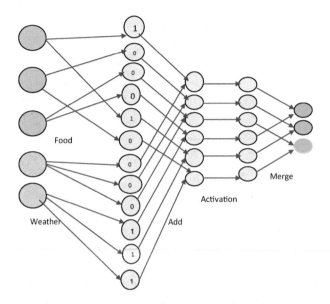

During step 2, data is added and put in stage 3 layer. This is

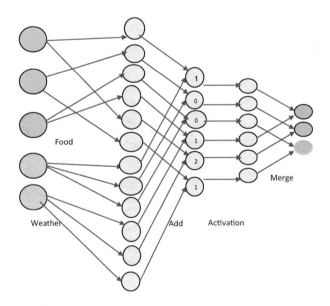

The activation function provides an output 1 for the maximum and sets the rest of the cells to zero. This is given by this figure:

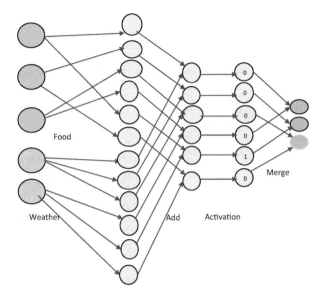

The last stage of the processing is merging the results into the output layer, which results in the diagram:

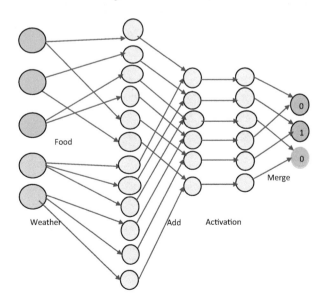

After this stage, data in the output layer is fed back to the input food layer to repeat the processing. The weather section is updated with the weather as it is today (which could as well be different or the same with the weather for the previous day). This step is

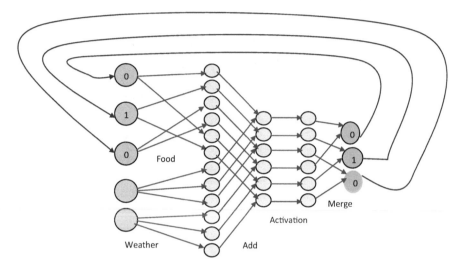

This completes the processing and the network is ready to repeat processing what should be the food to be cooked for the day. This depends on the weather for the day.

11

Convolutional Neural Networks

11.1 Introduction

As the application of neural networks extends beyond its legacy areas of image and pattern recognition, voice and speaker recognition and prediction of stock performance and prices, neural network concepts have also kept pace with the rapid developments in data analysis. This has gradually shifted the use of neural networks to beyond its traditional areas to also sit firmly within the gamut of applied data analytics and deep learning. In doing so, techniques which used to dominate signal processing such as filtering have been extended to the study of neural networks leading to convolutional neural networks (CNN). Convolutional neural network combines filter theory with pattern recognition concepts to provide new directions in signal analysis. In this chapter, emphasis will be placed on convolution neural networks. CNN applies filter theory, matrix concepts, convolution theory and neural networks. In doing so, it provides a basis for deep learning in a manner which makes computation of CNN applications fast and efficient.

11.2 Convolution Matrices

A convolution matrix is a data moving average window for interrogating large data sets and reducing their dimensions to more manageable levels. Convolution matrices are also filter coefficients arranged in a manner, which allows not only windowing of the data but also convolution of the filters with the data set. They reduce the traditional convolution in the time domain into direct multiplication of the window elements with data elements. In real sense, it can be argued that what is called convolution in convolution neural networks is not really what is well known in signal processing as convolution. In the time domain convolution is a wrapping of two signals around each other and to decouple the two signals, transformation of the two signals into the frequency domain is used. This is because convolution in the frequency domain reduces to multiplication of the Fourier transforms of the two signals.

Convolution matrix is a small matrix used in machine learning for feature detection and extraction. They help to zero in to most interest parts of a signal, where most interesting means the patterns hitherto buried in the signal that are made visible by the convolution matrices.

In digital signal processing parlance, discrete convolution of an input to a neural network and weight is defined for the one-dimensional (1D) case by the Equation (11.1)

$$y[i] = x[i] \times h[n] = \sum_{k=-\infty}^{\infty} x[k] \times h[i-k], \quad k \in [-\infty, \infty] \quad (11.1)$$

The convolution operator runs from minus infinity to plus infinity. In practice, this range is truncated to reasonable length K over which the operation is undertaken. In the rest of the chapter, the convolution expression will be given in terms of well defined signal lengths. This is done in the next expression for the case of two-dimensional convolution, which is defined by the expression:

$$y[i, j] = x \times w = \sum_{m=-M}^{M} \sum_{n=-N}^{N} x[i+m, j+n]w[m, n] \quad (11.2)$$

where, for example, the

- input is an image a two-dimensional (2D) array x containing pixels;
- convolution kernel or operator is also a 2D array of learnable parameters w;
- the output of the convolution is a 2D feature map y.

The convolution can be extended to more than two dimensions. For example, in a colour image containing red (R), green (G) and blue (B) or RGB planes, the convolution is a three-dimensional operation.

$$y[i, j, k] = x \times w = \sum_{m=-M}^{M} \sum_{n=-N}^{N} \sum_{l=-L}^{L} x[i+m, j+n, k+l]w[m, n, l]$$

$$(11.3)$$

The 3D convolution is popular in CNN applications involving 3D image processing and pattern recognition networks.

11.2.1 Three-Dimensional Convolution in CNN

The traditional convolution as given in the previous section involves shifts and the indices run from negative values to position values. In CNN, convolution is performed by direct multiplication of the convolution kernel with a section of the signal and summed before the kernel is moved as a unit and the process repeats. The convolution kernels are low-dimensional matrices, usually 3×3 and 5×5. For example, a typical 3D CNN convolution operation is of the form

$$y = \sum_{k=1}^{3}\sum_{i=1}^{5}\sum_{j=1}^{5} x[k,i,j]c[k,i,j] \qquad (11.4)$$

(3D versions)

where $c[k,i,j]$ is the convolution kernel or filter and $x[k,i,j]$ is a section of the 3D signal. The process is a direct multiplication (point by point) and summing.

During the training process in CNN, the kernel is adapted in response to the input signal.

11.3 Convolution Kernels

Convolution kernels have distinct functions. For example, in image processing, the convolution filters are used for sharpening, blurring, edge detection, edge enhancement, emboss and line detection. The choice of the convolution kernel to use is therefore very important in pattern recognition and classification. A few convolution kernels are given in this section:

Function of Kernel	Kernel (Filter) Matrix	Effect
Identity (leaves data unchanged)	$\begin{bmatrix} 0 & 0 & 0 \\ 0 & 1 & 0 \\ 0 & 0 & 0 \end{bmatrix}$	
Sharpen (image)	$\begin{bmatrix} 0 & 0 & 0 & 0 & 0 \\ 0 & 0 & -1 & 0 & 0 \\ 0 & -1 & 5 & -1 & 0 \\ 0 & 0 & -1 & 0 & 0 \\ 0 & 0 & 0 & 0 & 0 \end{bmatrix}$	

(Continued)

Continued

Function of Kernel	Kernel (Filter) Matrix	Effect
Blur	$\begin{bmatrix} 0 & 0 & 0 & 0 & 0 \\ 0 & 1 & 1 & 1 & 0 \\ 0 & 1 & 1 & 1 & 0 \\ 0 & 1 & 1 & 1 & 0 \\ 0 & 0 & 0 & 0 & 0 \end{bmatrix}$	
Blur (Gaussian)	$\dfrac{1}{16} \begin{bmatrix} 1 & 2 & 1 \\ 2 & 4 & 2 \\ 1 & 2 & 1 \end{bmatrix}$	
Edge Detector	$\begin{bmatrix} 0 & 1 & 0 \\ 1 & -4 & 1 \\ 0 & 1 & 0 \end{bmatrix}$	
	$\begin{bmatrix} -1 & -1 & -1 \\ -1 & 8 & -1 \\ -1 & -1 & -1 \end{bmatrix}$	
	$\begin{bmatrix} 1 & 0 & -1 \\ 0 & 0 & 0 \\ -1 & 0 & 1 \end{bmatrix}$	
	Sobel edge detectors: $G_x = \begin{bmatrix} -1 & 0 & 1 \\ -2 & 0 & 2 \\ -1 & 0 & 2 \end{bmatrix}$ $G_y = \begin{bmatrix} 1 & 2 & 1 \\ 0 & 0 & 0 \\ -1 & -2 & 1 \end{bmatrix}$	

Function of Kernel	Kernel (Filter) Matrix	Effect
Edge Enhancer	$\begin{bmatrix} 0 & 0 & 0 \\ -1 & 1 & 0 \\ 0 & 0 & 0 \end{bmatrix}$	
Emboss	$\begin{bmatrix} -2 & -1 & 0 \\ -1 & 1 & 1 \\ 0 & 1 & 2 \end{bmatrix}$	

Depending on the type of kernel used, edge detectors are known by various names, including Sobel, Prewit, Canny, Laplacian, Kirsch and Robinson and many more edge detectors. Many of the edge detectors are gradient-based. They are convolved with the image row and column wise. If the value of the pixel is more than the set threshold, it is taken as an edge.

Line detectors are used to detect lines of various orientations. The following kernels can detect horizontal, vertical and oblique (+45 and −45 degrees) lines.

$$\text{Horizontal line detector} = \begin{bmatrix} -1 & -1 & -1 \\ 2 & 2 & 2 \\ -1 & -1 & -1 \end{bmatrix}$$

$$\text{Vertical line detector} = \begin{bmatrix} -1 & 2 & -1 \\ -1 & 2 & -1 \\ -1 & 2 & -1 \end{bmatrix}$$

$$\text{Oblique (+45 degrees) line detector} = \begin{bmatrix} -1 & -1 & 2 \\ -1 & 2 & -1 \\ 2 & -1 & -1 \end{bmatrix}$$

$$\text{Oblique (−45 degrees) line detector} = \begin{bmatrix} 2 & -1 & -1 \\ -1 & 2 & -1 \\ -1 & -1 & 2 \end{bmatrix}$$

Convolutions are linear shift invariant, meaning that a linear shift operation does not alter the results of convolution. The complexity of a convolution operation depends on the sizes of the data and convolution kernel. Consider an image I of size M×N pixels and a convolution kernel C of size $(2R+1) \times (2R+1)$. Since the convolution operation is

$$I'(u, v) = \sum_{i=-R}^{R} \sum_{j=-R}^{R} I(u-1, v-j)C(i,j) \qquad (11.5)$$

The computation complexity of the convolution is of order $o(MN(2R+1)(2R+1)) \cong o(MNR^2)$. For a fixed sized image, the computation complexity is proportional to the order of the convolution kernel $o(R^2)$.

11.3.1 Design of Convolution Kernel

A simple approach for designing convolution kernels is to use separable design. In separable design, the kernel is a product of several terms. For example, a two-dimensional kernel could use a kernel which is a convolution of two terms of the form in Equation (11.6)

$$C = C_1 \times C_2 \qquad (11.6)$$

This approach thus uses a product of filters, one along the rows and the second along columns.

In general, an n-dimensional separable kernel can be designed with the expression (11.8)

$$C = C_1 \times C_2 \times \cdots \times C_n \qquad (11.7)$$

Seperability of a kernel is useful in reducing the complexity of computation by applying one component on one dimension and the second along the second dimension of the data.

Example: Consider the following separable kernels

$$C_x = \begin{bmatrix} 1 & 1 & 1 & 1 & 1 \end{bmatrix} \quad \text{and} \quad C_y = \begin{bmatrix} 1 \\ 1 \\ 1 \end{bmatrix}$$

The convolution kernel $C = C_x \times C_y = \begin{bmatrix} 1 & 1 & 1 & 1 & 1 \\ 1 & 1 & 1 & 1 & 1 \\ 1 & 1 & 1 & 1 & 1 \end{bmatrix}$

11.3.1.1 Separable Gaussian kernel

A typical example of a separable kernel is a Gaussian kernel formed from a Gaussian function of the form using first the one-dimensional Gaussian to design the two-dimensional Gaussian function

$$g_\sigma(x) = \frac{1}{\sqrt{2\pi}\sigma} \exp\left(-\frac{x^2}{2\sigma^2}\right) \tag{11.8}$$

and

$$G_\sigma(x,y) = \frac{1}{2\pi\sigma^2} \exp\left(-\frac{x^2+y^2}{2\sigma^2}\right) \tag{11.9}$$

The value σ determines the width of the Gaussian filter. The shape of the Gaussian function remains the same irrespective of the value taken by σ. In statistical parlance when the Gaussian distribution function is considered, σ is the standard deviation and σ^2 is the variance.

The two-dimensional result is because the two-dimensional Gaussian function is the product of two one-dimensional Gaussian functions.

$$
\begin{aligned}
G_\sigma(x,y) &= \frac{1}{2\pi\sigma^2} \exp\left(-\frac{x^2+y^2}{2\sigma^2}\right) \\
&= \frac{1}{\sqrt{2\pi}\sigma} \exp\left(-\frac{x^2}{2\sigma^2}\right) \times \frac{1}{\sqrt{2\pi}\sigma} \exp\left(-\frac{y^2}{2\sigma^2}\right) \\
&= g_\sigma(x) \times g_\sigma(y)
\end{aligned}
\tag{11.10}
$$

In general, the n-dimensional Gaussian kernel is

$$G_\sigma(x_1, x_2, \ldots, x_n) = \frac{1}{(\sqrt{2\pi}\sigma)^n} \exp\left(-\frac{x_1^2 + x_2^2 + \cdots + x_n^2}{2\sigma^2}\right) \tag{11.11}$$

The n-dimension kernel also maintains the Gaussian shape. The three-dimensional kernel is most suited to 3D convolution neural networks.

Example: Design a 3×3 Gaussian kernel using the following one-dimensional Gaussian filters

$$G_x = \begin{bmatrix} 1 & 2 & 1 \end{bmatrix} \quad \text{and} \quad G_y = \begin{bmatrix} 1 \\ 2 \\ 1 \end{bmatrix}$$

Solution: The 3×3 Gaussian filter is obtained by using the product $(3 \times 1 \times 1 \times 3)$

$$G = G_y G_x = \begin{bmatrix} 1 \\ 2 \\ 1 \end{bmatrix} \begin{bmatrix} 1 & 2 & 1 \end{bmatrix} = \begin{bmatrix} 1 & 2 & 1 \\ 2 & 4 & 2 \\ 1 & 2 & 1 \end{bmatrix}$$

11.3.1.2 Separable Sobel kernel

Apart from the Gaussian filter, the Sobel kernel is also obtained as a product of two one-dimensional filter kernels

$$S_y = \begin{bmatrix} 1 \\ 2 \\ 1 \end{bmatrix} \quad \text{and} \quad S_x = \begin{bmatrix} -1 & 0 & 1 \end{bmatrix}$$

$$G = G_y G_x = \begin{bmatrix} 1 \\ 2 \\ 1 \end{bmatrix} \begin{bmatrix} -1 & 0 & 1 \end{bmatrix} = \begin{bmatrix} -1 & 0 & 1 \\ -2 & 0 & 2 \\ -1 & 0 & 1 \end{bmatrix}$$

11.3.1.3 Computation advantage

The computational advantage in using separable filters depends on the dimensions of the filters. For the non-separable filters, we have shown that the computation complexity for filtering a 2D signal of size $\times N$ with a kernel of size P $\times$ Q is $o(MN(P)(Q)) \cong o(MNPQ)$. When the filter is separable, the computation order for each dimension is $o(MNP) + o(MNQ)$. Therefore, the computation advantage in using a separable filter is $G = MNPQ/(MNP + MNQ) = PQ/(P + Q)$. For example, for a 5 by 5 kernel the computation advantage is 2.5.

As a summary to this section, the convolution kernels also set the tone for the various functions of the convolution neural networks. For example, they provide the means for image retrieval, detection of objects within image sequences, types of objects in images, image classification and segmentation. In today's terms, they are also used in autonomous vehicles (self-driving cars, unmanned aerial vehicles), robots, face recognition, fingerprint recognition, pose and gait recognition. In the medical field, they are used for bioinformatic applications, for diagnosis of diseases and proteomics. In remote sensing, they are used for location recognition in aerial maps (streets, buildings and other objects within the environment), number plate recognition, captioning of images and object tracking.

11.4 Convolutional Neural Networks

From inception, CNN was developed with image analysis and pattern recognition in mind. Hence, most of the terminologies in use are related to image processing concepts in signal processing. This section describes the terminologies used in the description of convolution neural networks.

11.4.1 Concepts and Hyperparameters

Inputs to CNN are by far mostly image data with a wide range of variations in the dimensions of image frames and with frame rates 60 in the US and 50 frames per second in most parts of the world. Colour images also have various formats. In its simplest forms, colour images have three colour channels, red (R), green (G) and blue (B) planes leading to this form of images being referred to as RGB format. Each colour pixel therefore contains a triplet of bytes with each byte being 8 bits. Hence, an RGB pixel is 24 bits long. In each frame, a colour component in a pixel is therefore a value between (0 and 2^8 or 255). Hence, a colour image is a three-dimensional data set often referred to as input volume or $(3 \times 255 \times 255)$.

For a grey scale image, the input data is 2D and the data volume is basically a $1 \times 255 \times 255$ data channel. This greatly simplifies the operation of the convolutional neural network. For the rest of the discussion in this section, the focus is on the more difficult version of an RGB image input, which requires three image channels.

Hyperparameters refer to the properties of structure of CNN layers and neurons including the spatial arrangement and receptive fields. Hyperparameters are thus zero-padding (P), dimensions of the input volume (Width $\times$ Height $\times$ Depth = W $\times$ H $\times$ D), receptive field (R) and stride (S).

11.4.1.1 Depth (D)

Depth (D) refers to the number of filters used for the convolution operation. For each filter used, a feature map is produced. In an RGB scenario, three feature maps are produced each corresponding to the R, G and B components of the input image. Each feature map is a downsampled 2D version of the colour channel.

11.4.1.2 Zero-padding (P)

During convolution, the first row and last rows, first and last columns often have no means of having the convolution filter being centred on pixels in them. This exposes them to not be used optimally and lead to a reduction of

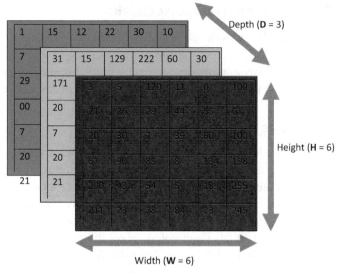

Figure 11.1   Input volume of size $6 \times 6 \times 3$ (Height $\times$ Width $\times$ Depth $= 6 \times 6 \times 3$) [1].

0	0	0	0	0	0	0	0
0	3	5	120	11	0	100	0
0	71	26	29	44	55	61	0
0	20	30	7	39	60	100	0
0	67	90	89	8	134	198	0
0	200	48	54	5	189	255	0
0	211	23	78	84	23	245	0
0	0	0	0	0	0	0	0

Figure 11.2   Zero padding of image channel.

the size of the activation maps. To enable application of the filter to the first element of a data matrix which do not have neighbouring elements to the top and left *zero-padding is used*. All the edge rows and column elements outside of the data matrix are made to be zero. Therefore, the filter can be applied to all the content of the input data matrix using zero-padding.

Zero padding has been a common operational process used in signal processing to reduce aliasing and edge effects. In convolutional neural networks, occasionally, it may be necessary to pad the edges of an image before undertaking the convolution. Several approaches are in use. Zero-padding (P) refers to when the edges of the input matrix is padded with zeros around the border. Zero-padding provides for centring of the filter matrix on each possible pixel of the input image. It also allows for control of the size of the feature maps. It is used in CNN layers when it is desired to preserve the dimensions of the output volume with the dimensions of the input volume. Use of zero-padding is also called *wide convolution*. Using no zero padding is called *narrow convolution*.

11.4.1.3 Receptive field (R)

Image inputs to CNN are usually of high dimensions. This makes it impractical to connect all possible regions of the image to neurons. Doing so will lead to very high computation cost or complexity. To reduce the computation cost associated with CNN operations, a small two-dimensional area of the data called the receptive field is used. Receptive fields are typically of about 5×5 or 3×3 or 7×7 matrix area for two-dimensional data. In a colour image, for example, it is necessary to extend the receptive field into the three colour channels so that it becomes either $5 \times 5 \times 3$ or other similar 3D constructs. It is assumed that all the pixels within the receptive field are thus fully connected to the CNN input layer. Over the receptive field, the network operates to create the activation map.

11.4.1.4 Stride (S)

The number of pixels by which the convolution kernel skips before it is placed again on the image to perform convolution is called the ***stride (S)***. It is the number of pixels by which the filter matrix is slid across the image. A stride of 1 means the filter matrix is moved by one pixel. When the stride is 2, the filter matrix skips two pixels to be placed on the image. Larger strides result to smaller feature maps. The value obtained may be normalised by the sum of the activation matrix.

11.4.1.5 Activation function using rectified linear unit

Use of activation functions is not new in neural network operations. While some activation functions perform better than others, it is always necessary to use them to produce outputs from neurons. ReLU is described in a previous chapter. ReLUs are used in CNN after every convolution step as a non-linear operation on the results of the convolution process. They are used to transform the output of convolution to binary values where a 1 refers to the neuron firing and a zero to it not firing. An element by element operation is undertaken with ReLU. For example, all negative outputs of the convolution operation are set to zero and all positive values to 1.

11.4.2 CNN Processing Stages

Four main processing operations are involved when convolution neural networks are used. They are:

(1) Convolution of the filter kernel with the input data.
(2) Application of a non-linearity (activation function). The ReLU or recti-fied linear unit is most popular with CNN. This step converts the output of the convolution operation to binary values.
(3) Pooling or subsampling of the input data. This step reduces the dimen-sionality of the data.
(4) Classification operation which creates a fully connected layer.

In the next sections, these four operations are described in details with examples and illustrative diagrams.

11.4.2.1 Convolution layer

The convolution layer is used as it preserves the spatial structure of the image. It is a segment of three image channels stacked on top of each other with each channel representing a colour plane in the RGB image format. A typical size would be of the form $32 \times 32 \times 3$, where the 3 in this number represents the three image channels. Each channel is of size 32 pixels by 32 pixels.

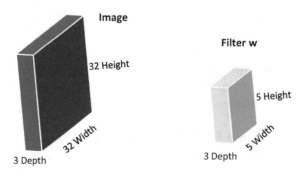

The convolution process uses, for example, a $5 \times 5 \times 3$ convolution kernel shown in the figure. We slide this over the filter and undertake a dot product of the filter with the image. This procedure is shown in the following figure. The result of the dot product when the filter is placed at a location on the image is a value.

This operation is written as:

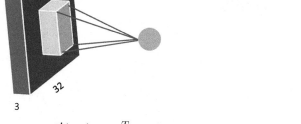

$$I'(i, j) = w^T x + b \qquad (11.12)$$

where x is the image window and b is the bias. For the kernel shown, the single value resulting from the operation requires a $5 \times 5 \times 3$ dot product. The kernel is slide across and down the image and for each location a value is obtained and recorded in a different image. The convolution is equivalent to subsampling of the image and adding the result to a new subsampled image. The process is called *pooling*.

Interestingly, the activation map is a two-dimensional layer of connected neurons providing outputs from the local areas exposed by the filter kernel. Each neuron is connected to the region. The neurons also share parameters.

By starting this process at a corner of the input image and sliding the convolution kernel across and centring it on each pixel, a new image or feature is obtained. This new image called *activation map* has lower dimensions compared with the input due to not being able to completely fit the kernel at the edges of the input image. Each activation map has a depth of 1. The stack of activation maps in Figure 11.3 is obtained using 6 filter kernels.

One of the main advantages of using this form of convolution with kernels is that different types of kernels may be used. Each different kernel picks a different kind of feature from the image. The subsampled images are stacked as shown and the process of convolution may be repeated on them as well.

Features are important or useful information or patterns derived from the input image by using the convolution kernel. In other words, the convolution neural network uses the convolution kernels to learn the features from the input image.

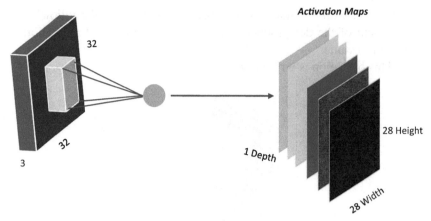

Figure 11.3   Activation maps in CNN.

These features are characteristic information on objects detected, human faces from the images, distinct objects, curves and lines picked up from the input image.

Observe that the convolved value obtained by summing the result of the dot product is a single entry in the activation matrix. The sliding is done over a stride of 1, or 2 or more as desired for the application.

Pooling

Consider the following convolution operation using a portion of the red channel of our data with emboss convolution kernel. No zero-padding is added to the edges of the data.

Example 1

3	5	120	11	0
71	26	29	44	55
20	30	7	39	60
67	90	89	8	134
200	48	54	5	189

$$x \begin{bmatrix} -2 & -1 & 0 \\ -1 & 1 & 1 \\ 0 & 1 & 2 \end{bmatrix} = \begin{bmatrix} 17 & 2 & -22 \\ 117 & 40 & 266 \\ 198 & 4 & 383 \end{bmatrix}$$

The sum of the convolution kernel is 1. Therefore, entries in the activation map are divided by 1 and remain unchanged.

The first entry of the result comes from the product: $[3 \times (-2) + (-1) \times 5 + 0 - 71 \times 1] + [26 \times 1 + 29 \times 1 + 0] + [0 + 30 \times 1 + 7 \times 2 = (-6 - 5 - 71) + (26 + 29 + 30 + 14) = -82 + 99 = +17$.

The remaining convolution values are computed the same way using the dot product of the corresponding elements of the kernel and the data window or receptive field.

We can always tell what the size activation map would be using the following expression Given the stride S, and the dimension of the kernel or filter F, if the dimension of the square image is N, then the size of the output activation map is:

$$M = \frac{(N - F)}{S} + 1 \qquad (11.13)$$

In other words, the activation map has size M × M. When the input image is padded, the new dimension N in the above equation should include the padding and indeed N has increased by 2 giving a larger activation map. One more hidden information in the above equation is that the size or number of padding rows and columns used for a stride $S = 1$ is given generally by $(F - 1)/2$. Using this padding size restores the size of the activation map to the size of the input image. When we pad by p rows (column), we can modify the above equation to be:

$$M = \frac{(N + 2P - F)}{S} + 1 \qquad (11.14)$$

11.4.3 The Pooling Layer

The pooling layer is positioned after the convolutional layer. It is used primarily to reduce the spatial dimensions (Width × Height) of the Input Volume for the next Convolutional Layer. It has no effect on the depth dimension of the volume.

The role of the pooling layer is two-fold, to perform downsampling of the activation map thereby making them more manageable for the next repeat operations of convolution. When several activation maps are stacked as in Figure 11.3, the pooling layer operates on each activation map independently. Downsampling usually leads to loss of information that is not recoverable. There are signal processing techniques, which use quadrature mirror filters for down sampling, which makes it possible to recover the input data without loss of information. In signal processing that requires the use of an analysis filter for downsampling and synthesis filter for recombining the sampled components. There appears to be no equivalent operations in CNN to date. However, losses in convolutional networks have inherent benefits, which include:

(1) decrease in computational overhead for the next sets of layers of the network;

(2) protection against over-fitting.

The pooling operation performs an operation similar with convolution by using a sliding window. The sliding window is moved in strides to transform the activation map into feature. Several types of kernels are popular. Among them are kernels which extract the maximum value from the values exposed by the sliding window. This is called 'max pooling'. Another approach is by taking the average of the values. Max pooling provides better performance over other approaches and for that reason has become the favourite approach.

For each depth slice, pooling operation is undertaken. For example, for an input volume of size W × D a pooling window of size 2 × 2 is used on each colour slice. This means that the colour channels are equivalently downsampled.

Pooling is optional in convolutional neural networks. However, it can prove very useful if, for example, the colour channels are very big making downsampling necessary to speed up the operations of CNNs.

Max Pooling

In this operation, the kernel takes the maximum value from the activation map is used. Zero padding is avoided at this level as it adds edge effects to the outputs. Stride values of 1 and 2 with F = 2 or 3 are common for pooling operations. In a larger activation map, larger strides could be used as well.

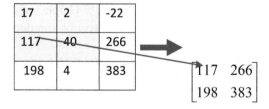

Notice that the 2×2 kernel selects only the maximum value within the receptive field. This example demonstrates that the resulting activation map or pooling layer depends strongly on the kernel or filter used for convolution. It makes sense that max pooling performs better than other methods. When

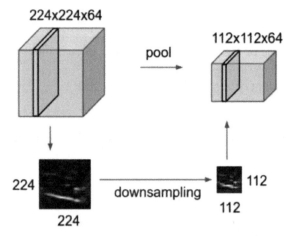

Figure 11.4 Example of pooling operation in CNN [1].

neurons fire based on a filter, the maximum value is a better representation of how strongly the neuron has fired compared with the smaller values.

11.4.4 The Fully Connected Layer

The fully connected layer is fully connected with the output of the previous layer. The term implies that the neurons in the previous layer is connected to all the neurons in the next layer. The fully connected layer contains neurons which are fully connected to the input volume. They are used typically in the last stage of the convolution neural networks and to connect to the output layer to construct the desired number of outputs.

11.5 CNN Design Principles

This section is a summary of the design of convolution neural networks. Relevant expressions at each layer of the processing steps are given.

Step 1

The input to the CNN is defined by the input volume, a product of the width, height and depth of the input data with the expression

$$V_1 = W_1 \times H_1 \times D_1 \tag{11.15}$$

The subscript '1' specifies the input layer 1. Processing the input volume requires four parameters given by:

(1) The number of filters used K
(2) The spatial dimension of the filter F
(3) The stride or the steps over which the filter is moved across the data S and
(4) The amount of zero padding (when used) P

Step 2

The above parameters are applied to the input volume to create a new volume with

Size

$$V_2 = W_2 \times H_2 \times D_2 \qquad (11.16)$$

The width and heights for the second layer are re-definitions of the same parameters for the first layer. This occurs as the outcome of processing of the first layer. Generally, if the data volume is padded, the expressions for the parameters are:

$$\left. \begin{aligned} W_2 &= \left(\frac{W_1 - F + 2P}{S} \right) + 1 \\ H_2 &= \left(\frac{H_1 - F + 2P}{S} \right) + 1 \\ D_2 &= K \end{aligned} \right\} \qquad (11.17)$$

Per filter the number of weights is found to be $F \cdot F \cdot D_1 K$. The number of biases is K.

Common Sizes

To ease processing, the values of K are chosen as powers of 2. Typical values include (32, 64, 128, 256 and 512). The values of F the filter dimension, S and P are typically

(1) $F = 1, S = 1, P = 0$ (no padding)
(2) $F = 3, S = 1, P = 1$
(3) $F = 5, S = 1, P = 2$
(4) $F = 5, S = 2, P = 2$ or 1

11.6 Conclusion

This chapter is an introduction of convolution neural networks. Filter kernels were introduced to be used in convolution operations involving images. Different kernels perform different roles such as edge detection, image enhancement and embossing. These techniques dominate signal processing. They have been extended to the study of neural networks leading to convolutional neural networks (CNN). Thus convolutional neural network combines filter theory with pattern recognition concepts to provide new directions in signal analysis. Emphasis is placed on convolution neural networks. CNN filter theory, matrix concepts, convolution theory and neural networks were analyzed. They provide insights into deep learning in a manner which makes computation of CNN applications fast and efficient.

Reference

[1] Fei-Fei Li, Justin Johnson and Serena Yeung, "Lecture 5 | Convolutional Neural Networks" on YouTube, Stanford University lecture notes.

12

Principal Component Analysis

12.1 Introduction

Principal component analysis is a method of determining hidden patterns in data. It has found use in many applications including meteorological data, face recognition, facial expression recognition and detection of emotions [1].

12.2 Definitions

Principal component is the direction of maximum variability (covariance) in the data set. This is called the first principal component. The second principal component is the next orthogonal or uncorrelated direction for greatest variation of the data. Thus, it is essential to first extract the variability along the principal component and to find the direction of next greatest variability. Therefore, once the principal component is removed, the second principal component points to the direction of maximum variation of the remaining or residual data sub-space. This component is orthogonal to the principal component. By inference, the third principal component is the principal component of the residual data sub-space after the second principal component is removed. These directions are always orthogonal to each other. They are determined by the values of the variance of the data.

By inference, therefore, it is essential to calculate the covariance of the data to find values of variances among the data set and hence enable finding or removal of the principal components.

There are several approaches to explaining what PCA is. All of them are explanations of the same concept using different terminologies.

From the projection mathematics point of view, PCA is a linear projection of data aimed at reducing the number of parameters or dimensionality of the data. The projections can thus be chosen to minimize square errors. The projection of a vector $\mathbf{v}$ to the axis $\mathbf{x}$ is given by $p = \mathbf{v} \cdot \mathbf{x} = vx \cos \theta$ and θ is

205

the angle between the two vectors. The direction with the greatest variation is determined by $E((\mathbf{v} \cdot \mathbf{x})^2)$. The expectation is maximum when the angle between the vector $\mathbf{v}$ and $\mathbf{x}$ is zero. In other words, the vector lies along the axis $\mathbf{x}$.

In practice, the data is mapped into a lower dimensionality and therefore may be viewed as a form of unsupervised learning. The algorithm transforms a set of correlated data and decouples them into uncorrelated data sets. This is essential to examine each data and its influence on the process. Therefore, for example, given an N-dimensional data, PCA can be used to reduce the dimension of the data to K-dimensions, where K≪N [1]. This is illustrated in Equation (12.1). Thus, a 3D data is mapped into a planar or 2D data; a 5D data is mapped, for example, into a 3D data. This is useful, for example, when we have N sensors each providing readings and the objective is to reduce the data set to smaller dimensions less than N.

$$
x = \begin{bmatrix} a_1 \\ a_2 \\ \vdots \\ a_N \end{bmatrix} \rightarrow \text{ reduce dimensionality of data } \rightarrow y = \begin{bmatrix} b_1 \\ b_2 \\ \vdots \\ b_K \end{bmatrix} \quad K \ll N
$$

$$(12.1)$$

Resulting from reduction in dimensionality, PCA also establishes directions of largest variability of data. In doing so, it rotates the existing axes into new positions in the space covered or defined by the data variables. The new axes defined by PCA are orthogonal to each other in the directions of largest variations for each data sub-space. Orthogonality also establishes separability of data.

PCA therefore produces at least three types of information. It produces

(i) orthogonal functions which reveal the pattern or structure of information in the data

(ii) principal components which are the time series that reveal the contributions from each of the orthogonal functions at a point in time. Thus, when the principal component PC_j is large, the contribution of the orthogonal function j is also large at the point in time. It also means that a small principal component PC_j shows small contributions from the orthogonal function j as well.

(iii) eigenvalues which provide the relative importance of the orthogonal functions.

Linear Transformation

To reduce data dimensionality, PCA uses a linear transformation T in the form

$$y = Tx \tag{12.2}$$

where the transformation T can be written in matrix form. The matrix represents a system of linear expressions relating x to y as

$$T = \begin{bmatrix} t_{11} & t_{12} & \cdots & t_{1N} \\ t_{21} & t_{22} & \cdots & t_{2N} \\ \cdots & \cdots & \ddots & \vdots \\ t_{K1} & t_{K2} & \cdots & t_{KN} \end{bmatrix} \tag{12.3}$$

By using Equations (12.1) and (12.2) to substitute for x and T, the linear transformation between x and y can be seen clearly to be the system of simultaneous equations in (12.4).

$$\left. \begin{array}{l} b_1 = t_{11}a_1 + t_{12}a_2 + \cdots + t_{1N}a_N \\ b_2 = t_{21}a_1 + t_{22}a_2 + \cdots + t_{2N}a_N \\ \vdots \qquad \cdots \\ b_K = t_{K1}a_1 + t_{K2}a_2 + \cdots + t_{KN}a_N \end{array} \right\} \tag{12.4}$$

An approach for achieving this transformation uses a set of basis functions in the high-dimensional space to represent x. A similar basis is also used in the reduced space to represent y. Semantically, this may be written in terms of the basis in the two spaces as

$$x = a_1 v_1 + a_2 v_2 + \cdots + a_N v_N \tag{12.5a}$$

The set $v_1, v_2, \ldots, v_N$ is a basis of the N-dimensional space [1]. Similarly, in the reduced K-dimensional space the signal y is representable in terms of a basis as

$$y = \hat{x} = b_1 u_1 + b_2 u_2 + \cdots + b_K u_K \tag{12.5b}$$

The set $u_1, u_2, \ldots, u_K$ is a basis of the K-dimensional space and $y = \hat{x}$ is a reasonable and suitable approximation of the data x. When $K = N, y = x$ and there is no reduction in the data set. The approximation in equation (12.5b) means that there are losses, which is not recoverable. The objective of PCA is however to minimize the losses as much as possible. This is represented with the error norm $e = \|x - \hat{x}\|$.

The rest of the chapter is dedicated to how to choose the bases to keep the error norm to a minimum so that the reduced dimensional subspace is chosen optimally. The reduced dimension space is determined through the choice of 'best' eigenvectors of the covariance matrix of x. In other words, eigenvectors corresponding the largest eigenvalues (called *principal components*) are selected to determine the subspace. The implied linear transformation is of the form

$$
\begin{bmatrix} b_1 \\ b_2 \\ \vdots \\ b_K \end{bmatrix} = \begin{bmatrix} u_1^T \\ u_2^T \\ \vdots \\ u_K^T \end{bmatrix} \times (x - \hat{x}) = U^T (x - \hat{x}) \tag{12.6}
$$

The criteria for choosing the best eigenvalues use thresholds of the form

$$
\frac{\sum_{j=1}^{K} \lambda_j}{\sum_{j=1}^{N} \lambda_j} > \ell \tag{12.7}
$$

where ℓ is the threshold, for example, 90% (or 95%). Choosing a threshold of 90% means that 90% of the information in the data is preserved and a loss of 10%. Using a higher threshold like 95% reduces the amount of information loss to about 5%.

Reconstruction of Data from PCA

A good and representative version of the original data can be reconstructed from the approximation in the reduced dimensional space by using the expression

$$
\hat{x} - \bar{x} = \sum_{i=1}^{K} b_i u_i \tag{12.8a}
$$

or

$$
\hat{x} = \sum_{i=1}^{K} b_i u_i + \bar{x} \tag{12.8b}
$$

where $\bar{x}$ is the mean of the input data and $\hat{x}$ is the reconstructed data. The reconstruction error is thus $e = \|x - \hat{x}\|$. This error can be shown to be equal to

$$
e = \frac{1}{2} \sum_{K+1}^{N} \lambda_i \tag{12.9}
$$

The error is proportional to the sum of rejected or smaller eigenvalues.

12.2.1 Covariance Matrices

The mathematical foundations for PCA depend strongly on understanding of what a covariance matrix is and what eigenvalues and eigenvectors are. This chapter therefore will first provide a basis for understanding these concepts and then apply them directly to practical data.

Consider the following environmental data X containing readings from six sensors recording the concentration of ethanol at a location.

S1_max	S2_max	S3_max	S4_max	S5_max	S6_max
0.09353	0.09985	0.08644	0.38398	0.12024	0.05778
0.0808	0.08781	0.14446	0.28284	0.09405	0.05509
0.10897	0.10989	0.16347	0.36844	0.1299	0.06762
0.113	0.10598	0.09914	0.38801	0.1395	0.06674
0.15199	0.15098	0.20885	0.44514	0.17912	0.09931
0.14486	0.14459	0.13876	0.38683	0.16964	0.10333

The correlation between these readings can be estimated for the sake of using it to estimate the principal components for these readings.

Several variables are essential for computing PCA algorithms. These are the mean or average of the data, the variance (the so-called energy inherent in the data) and standard deviation or the square root of the variance. Both the variance and standard deviation thus provide measures of the spread of the data or variability of the data. To initiate computing the variance, the mean of the data is first removed from all data samples. This allows the variance to be correctly interpreted as the variation from the mean of the data.

In an earlier chapter, covariance matrices were introduced and used in Kalman filter operations. They are used to compare statistical data to establish the levels of similarity and relationships. The following definitions were presented for data analysis.

x_i is individual measurements;
$\bar{X}$ is the mean or average of the measurements;
$(x_i - \bar{X})$ is the deviation from the mean so that it takes either positive or negative values which depends on the difference;
$(x_i - \bar{X})^2$ is the square of the deviation and is a positive number.

The variance of the data sequence $\{X\}$ of N samples is therefore computed with the equation (12.10).

$$\sigma_x^2 = \frac{\sum_{i=1}^{N} (x_i - \bar{X})^2}{N} \tag{12.10}$$

Standard deviation is the square root of variance and is given by the equation (12.11).

$$\sigma_x = \sqrt{\frac{\sum_{i=1}^{N}(x_i - \bar{X})^2}{N}} \tag{12.11}$$

The variance is a measure of how the data sequence lies around the mean value. The variance is a positive number. Suppose there is a second data sequence {Y} of mean $\bar{Y}$ also of the same length at the sequence {X}. The covariance of the two data sequences is given by equation (12.12).

$$\sigma_x \sigma_y = \frac{\sum_{i=1}^{N}(x_i - \bar{X})(y_i - \bar{Y})}{N} \tag{12.12}$$

Observe that the mean for each data type is subtracted first from the data samples in the calculation of the covariance of the data. The covariance is the product of the two standard deviations for X and Y data sequences. With the above standard definitions, the covariance matrices for one-, two- and three-dimensional Kalman filter are defined by the following equation (12.13):
One dimension (1D)

$$\rho = \sigma_x^2 = \left[\frac{\sum_{i=1}^{N}(x_i - \bar{X})^2}{N}\right] \tag{12.13}$$

The covariance is a measure of the energy in the signal and is useful in applications where the signal-to-noise ratio of a signal need to be determined.

Two Dimensions (2D)

For two dimensions, the covariance matrix is a 2×2 matrix given by the Equation (12.14)

$$\rho = \begin{bmatrix} \sigma_x^2 & \sigma_x \sigma_y \\ \sigma_x \sigma_y & \sigma_y^2 \end{bmatrix}$$

$$= \begin{bmatrix} \dfrac{\sum_{i=1}^{N}(x_i - \bar{X})^2}{N} & \dfrac{\sum_{i=1}^{N}(x_i - \bar{X})(y_i - \bar{Y})}{N} \\ \dfrac{\sum_{i=1}^{N}(y_i - \bar{Y})(x_i - \bar{X})}{N} & \dfrac{\sum_{i=1}^{N}(y_i - \bar{Y})^2}{N} \end{bmatrix} \tag{12.14}$$

Exercise 1

The following data represent readings from two sensors measuring the concentration of ethanol in an environment.

S2_max	S3_max
0.09985	0.08644
0.08781	0.14446
0.10989	0.16347
0.10598	0.09914
0.15098	0.20885
0.14459	0.13876

Calculate the covariance matrix for the density of ethanol in the room.

Exercise 2: The following data represents temperature and humidity recordings by two sensors located along the river Nile. Compute the covariance matrix for the data set.

X = Temp.	Y = Humidity
26.1349	58.9679
26.1344	58.9679
26.134	58.9679
26.1336	58.9678
26.1333	58.9678
26.133	58.9678
26.1327	58.9678
26.1326	58.9678
26.1333	58.9677
26.1339	58.9677
26.1345	58.9677
26.135	58.9677
26.1355	58.9677
26.1359	58.9677
26.1363	58.9677
26.1367	58.9677
26.137	58.9677
26.1373	58.9676
26.1375	58.9676

Three Dimensions (3D)

In three dimensions, the covariance matrix is given by Equation (12.15):

$$\rho = \begin{bmatrix} \sigma_x^2 & \sigma_x\sigma_y & \sigma_x\sigma_z \\ \sigma_y\sigma_x & \sigma_y^2 & \sigma_y\sigma_z \\ \sigma_z\sigma_x & \sigma_z\sigma_y & \sigma_z^2 \end{bmatrix}$$

$$= \begin{bmatrix} \dfrac{\sum_{i=1}^{N}(x_i-\bar{X})^2}{N} & \dfrac{\sum_{i=1}^{N}(x_i-\bar{X})(y_i-\bar{Y})}{N} & \dfrac{\sum_{i=1}^{N}(x_i-\bar{X})(z_i-\bar{Z})}{N} \\ \dfrac{\sum_{i=1}^{N}(y_i-\bar{Y})(x_i-\bar{X})}{N} & \dfrac{\sum_{i=1}^{N}(y_i-\bar{Y})^2}{N} & \dfrac{\sum_{i=1}^{N}(y_i-\bar{Y})(z_i-\bar{Z})}{N} \\ \dfrac{\sum_{i=1}^{N}(z_i-\bar{Z})(x_i-\bar{X})}{N} & \dfrac{\sum_{i=1}^{N}(z_i-\bar{Z})(y_i-\bar{Y})}{N} & \dfrac{\sum_{i=1}^{N}(z_i-\bar{Z})^2}{N} \end{bmatrix}$$

$$(12.15)$$

The standard deviation of the data provides a means of assessing the nature of the distribution of a data sequence. Normally, about 68% of all the measurements lie within $(\pm\sigma)$ one standard deviation from the mean. All measurement samples also lie within $\pm\sigma^2$ from the mean value.

The expression for covariance matrix for the multi-dimensional case may be derived from the expressions for the one, two and three dimensions as

$$\rho = \begin{bmatrix} \sigma_x^2 & \sigma_x\sigma_y & \sigma_x\sigma_z & \cdots & \sigma_x\sigma_t \\ \sigma_y\sigma_x & \sigma_y^2 & \sigma_y\sigma_z & \cdots & \sigma_y\sigma_t \\ \sigma_z\sigma_x & \sigma_z\sigma_y & \sigma_z^2 & \cdots & \sigma_z\sigma_t \\ \vdots & \vdots & \vdots & \ddots & \vdots \\ \sigma_t\sigma_x & \sigma_t\sigma_y & \sigma_t\sigma_z & \cdots & \sigma_t^2 \end{bmatrix} \qquad (12.16)$$

Many data processing applications which involve covariance matrices require the use of multi-dimensional covariance matrices. The more the number of independent variables, the larger the matrices and the more complex the processing of the covariance matrices.

Interpreting the covariance properly is as important as computing them correctly. The sign of the covariance value is particularly important.

A negative value of covariance means that when one value increases the other decreases. This could be seen intuitively, for example, when a student's absence in class is high, his grades also go down and when absence from classes is low, his grades go up.

A positive covariance value shows that both variables or dimensions increase or decrease jointly. For example, high class attendance also results to higher grades in subjects studied.

Zero covariance values mean there is no relationship between the two variables. They are independent of each other. For example, a student's height and marks scored in various subjects are independent of each other. While some of these conclusions can be drawn from some data, others need the scientist to go further into computing the eigenvalues and hence principal components to reveal hidden structures or truths in the data.

12.3 Computation of Principal Components

Several steps are involved in computing the principal components of a data set. They are described and explained in detail in this section. Three methods will be presented using vector analysis, covariance matrices and singular value decomposition.

12.3.1 PCA Using Vector Projection

The analysis use vector analysis to explain what is happening. Vectors are used intentionally to inform the reader on the fact that projections are involved and that orthogonal axes are also described as vectors. The projection of a vector $\mathbf{x}$ to the axis $\mathbf{u}$ is given by $p = \mathbf{u} \cdot \mathbf{x} = ux \cos \theta$ and θ is the angle between the two vectors. The direction with the greatest variation is determined when $E((\mathbf{u} \cdot \mathbf{x})^2)$.

Let
$$E(p^2) = E((\mathbf{u} \cdot \mathbf{x})^2) \tag{12.17}$$

Rewrite this expression by first expanding it through substitution leading to Equation (12.18):

$$\mathrm{E}((\mathbf{u} \cdot \mathbf{x})^2) = \mathrm{E}((\mathbf{u} \cdot \mathbf{x})(\mathbf{u} \cdot \mathbf{x})^{\mathrm{T}}) = \mathrm{E}(\mathbf{u} \cdot \mathbf{x} \cdot \mathbf{x}^{\mathrm{T}} \cdot \mathbf{u}^{\mathrm{T}}) \tag{12.18}$$

Define the matrix C as the correlation matrix for the vector x as in Equation (12.19).

$$C = \mathbf{x} \cdot \mathbf{x}^{\mathbf{T}} \tag{12.19}$$

Let $\mathbf{u}$ be a unit vector along an axis. Define the quantity w to maximize the product $\mathbf{u}\mathbf{C}\mathbf{u}^{\mathbf{T}}$. The quantity w may be found when it is the principal eigenvector of the matrix C. Therefore, we can write

$$\mathbf{u}\mathbf{C}\mathbf{u}^{\mathrm{T}} = \mathbf{u}\lambda\mathbf{u}^{\mathrm{T}} = \lambda \tag{12.20}$$

where λ is the principal eigenvalue of the matrix C. It represents the variability along the direction of the principal component.

To find the eigenvalue, the Lagrangian of the process is used. This involves maximizing the quantity $\mathbf{u^T xx^T u}$ subject to the inner product $\mathbf{u^T U} = 1$. The Lagrangian is Equation (12.21):

$$u^T xx^T u - \lambda u^T u \tag{12.21}$$

Giving the vector of partial derivatives with respect to the transpose of the unit vector **u** as

$$xx^T u - \lambda u = (xx^T - \lambda)u = 0$$

Since the unit vector u is not zero $(\mathbf{u} \neq 0)$, it must be an eigenvector of the matrix $\mathbf{C = xx^T}$ and has a value λ.

12.3.2 PCA Computation Using Covariance Matrices

Let the sequence of data be $x = \{x_1, \ldots, x_M\}$

(i) Compute the mean of the data. The mean is

$$\bar{x} = \frac{1}{M} \sum_{i=1}^{M} x_i$$

(ii) Subtract the mean from each value of the data and compute the covariance matrix as

$$\sum = \frac{1}{M} \sum_{i=1}^{M} (x_i - \bar{x})(x_i - \bar{x})^T$$

(iii) Compute the eigenvalues and eigenvectors with the expression

$$\sum \cdot v = \lambda v$$
$$\left(\sum \cdot - \lambda I \right) v = 0$$

where I is identity matrix of the same order as the covariance matrix. Since the vector $v \neq 0$ the quantity

$$\sum - \lambda I = 0.$$

The solution of the determinant of this expression gives the eigenvalues of the covariance matrix. For small matrices $\sum$, simultaneous equations may be used to solve this equation. For each eigenvalue, there is a corresponding eigenvector.

(iv) To find the eigenvectors, we substitute the eigenvalues into the matrix

(v) Let $B = \sum -\lambda I$, then the eigenvector equation is $Bx = (\sum -\lambda_j I)x = \begin{bmatrix} 0 \\ \vdots \\ 0 \end{bmatrix}$. For each eigenvalue, this equation is solved to find the corresponding eigenvector.

For the m-th order equation, there are at most m eigenvalues and m eigenvectors.

Covariance matrices are generally symmetric in nature. For symmetric matrices, the eigenvectors are orthogonal. Hence, for any pair of eigenvalues of a symmetric matrix λ_i and λ_j, the inner product (dot product) of their eigenvectors is zero, meaning that their eigenvectors v_i and v_j are orthogonal. Their inner product is

$$v_i \times v_j = 0$$

This also means that the following expression holds for all values of j:

$$\sum v_j = \lambda_j v_j; \quad \forall j$$

Example: Consider the real symmetric covariance matrix $\sum = \begin{bmatrix} 2 & 1 \\ 1 & 2 \end{bmatrix}$.

Calculate its eigenvalues and eigenvectors. Show that its eigenvectors are orthogonal.

$$\det\left(\sum -\lambda I\right) = \begin{vmatrix} 2 - \lambda & 1 \\ 1 & 2 - \lambda \end{vmatrix} \Rightarrow (2 - \lambda)^2 - 1 = 0.$$

The resulting quadratic equation has two roots $\lambda_1 = 1$ and $\lambda_2 = 3$.

The eigenvectors are given by the expression:

$$\begin{pmatrix} 2 - \lambda & 1 \\ 1 & 2 - \lambda \end{pmatrix} \begin{pmatrix} x_1 \\ x_2 \end{pmatrix} = \begin{pmatrix} 0 \\ 0 \end{pmatrix}$$

To find the eigenvector corresponding to the first eigenvalue, we have

$$\begin{pmatrix} 2 - 1 & 1 \\ 1 & 2 - 1 \end{pmatrix} \begin{pmatrix} x_1 \\ x_2 \end{pmatrix} = \begin{pmatrix} 0 \\ 0 \end{pmatrix}$$

and

$$x_1 + x_2 = 0$$

The two rows of the matrix equation result to the same linear equation. Choose a value for $x_1 = 1$; hence, $x_{21} = -1$ and the eigenvector corresponding to the first eigenvalue is $v_1 = \begin{bmatrix} 1, & -1 \end{bmatrix}$.

The eigenvector equation for the second eigenvalue is

$$\begin{pmatrix} 2-3 & 1 \\ 1 & 2-3 \end{pmatrix} \begin{pmatrix} x_1 \\ x_2 \end{pmatrix} = \begin{pmatrix} 0 \\ 0 \end{pmatrix} \Rightarrow \begin{pmatrix} -x_1 + x_2 = 0 \\ x_1 - x_2 = 0 \end{pmatrix}$$

From the first equation, $x_1 = x_2$. Hence, by choosing $x_1 = 1$ and $x_2 = 1$, the eigenvector corresponding to the second eigenvalue is $v_2 = \begin{bmatrix} 1, 1 \end{bmatrix}$.

To show that the two eigenvectors are orthogonal, we take their dot product which is

$$v_1 \cdot v_2 = \begin{bmatrix} 1 \\ 1 \end{bmatrix} \cdot \begin{bmatrix} 1 \\ -1 \end{bmatrix} = (1 \times 1) + (-1 \times 1) = 0$$

Example 2

Given the non-symmetric matrix $\Sigma = \begin{pmatrix} 0 & 1 \\ -2 & -3 \end{pmatrix}$, find its eigenvalues.

Solution

The eigenvalues are given by the equation

$$\Sigma - \lambda I = \begin{pmatrix} 0 & 1 \\ -2 & -3 \end{pmatrix} - \begin{pmatrix} \lambda & 0 \\ 0 & \lambda \end{pmatrix} = 0$$

This gives the matrix $\begin{pmatrix} -\lambda & 1 \\ -2 & -3-\lambda \end{pmatrix} = 0$. The determinant of this equation is

$$\det \begin{pmatrix} -\lambda & 1 \\ -2 & -3-\lambda \end{pmatrix} = \begin{vmatrix} -\lambda & 1 \\ -2 & -3-\lambda \end{vmatrix} = 0$$

$$-\lambda(-3-\lambda) + 2 = 0$$

$$\lambda^2 + 3\lambda + 2 = 0$$

The two eigenvalues are $\lambda_1 = -1$; $\lambda_2 = -2$. The bigger eigenvalue is λ_1. It is normal to order the eigenvalues in decreasing magnitude, and in this case, it is in the order $\lambda_1 \geq \lambda_2$. Provided the covariance matrix is a square matrix, we can find its eigenvectors.

Example 2

Using the solutions of Example 1, find the eigenvectors.

By substituting the first eigenvalue, we compute the corresponding eigenvector as

$$(1) \quad \begin{pmatrix} -\lambda & 1 \\ -2 & -3-\lambda \end{pmatrix} \begin{pmatrix} x_1 \\ x_2 \end{pmatrix} = \begin{pmatrix} 0 \\ 0 \end{pmatrix}$$

$$\begin{pmatrix} 1 & 1 \\ -2 & -3+1 \end{pmatrix} \begin{pmatrix} x_1 \\ x_2 \end{pmatrix} = \begin{pmatrix} 0 \\ 0 \end{pmatrix}$$

$$\begin{pmatrix} 1 & 1 \\ -2 & -2 \end{pmatrix} \begin{pmatrix} x_1 \\ x_2 \end{pmatrix} = \begin{pmatrix} 0 \\ 0 \end{pmatrix}$$

We have

$$x_1 + x_2 = 0$$
$$-x_1 - x_2 = 0$$
$$x_1 = -x_2$$

We can now pick values for the eigenvector components as $x_1 = 1$; $x_2 = -1$. The eigenvector corresponding to eigenvalue -1 is $v_1 = \begin{bmatrix} 1, & -1 \end{bmatrix}$.

(2) We can now solve for the second eigenvector by using the expression

$$\begin{pmatrix} -\lambda & 1 \\ -2 & -3-\lambda \end{pmatrix} \begin{pmatrix} x_1 \\ x_2 \end{pmatrix} = \begin{pmatrix} 0 \\ 0 \end{pmatrix}$$

$$\begin{pmatrix} 2 & 1 \\ -2 & -3+2 \end{pmatrix} \begin{pmatrix} x_1 \\ x_2 \end{pmatrix} = \begin{pmatrix} 0 \\ 0 \end{pmatrix}$$

$$\begin{pmatrix} 2 & 1 \\ -2 & -1 \end{pmatrix} \begin{pmatrix} x_1 \\ x_2 \end{pmatrix} = \begin{pmatrix} 0 \\ 0 \end{pmatrix}$$

$$2x_1 + x_2 = 0$$
$$-2x_1 - x_2 = 0$$
$$-2x_1 = x_2$$

Let $x_1 = 1$; $x_2 = -2$. The corresponding eigenvector is $v_2 = \begin{bmatrix} 1, -2 \end{bmatrix}$

12.3.3 PCA Using Singular-Value Decomposition

Singular-value decomposition (SVD) is a mathematical tool for decomposing (or factoring) matrices to products of matrices. Given a matrix A of size m×n, it can be factored into the following three matrices:

$$A = \cup W V^T$$

where

(a) U is an m × m matrix. The columns of U are eigenvectors of the matrix AA^T. U is orthonormal or $U^T U = I$.

(b) V is an n × n matrix. The columns of V are eigenvectors of the matrix $A^T A$. V is orthonormal so that $V^T V = VV^T = I$.

(c) W is an m × n matrix. W is a diagonal matrix containing the singular values of A such that $W = diag(\sigma_1, \sigma_2, \ldots, \sigma_r)$. The singular values are square roots of the eigenvalues $(\lambda_1, \lambda_2, \ldots, \lambda_r)$ of the matrices AA^T and of $A^T A$, given by the relationship $\sigma_j = \sqrt{\lambda_j}$.

(d) The rank of the matrix A is equal to the number of its non-zero singular values

(e) The matrix A is singular if at least one of its singular values is zero

For symmetric matrix A, there exists a unique decomposition such that

$$A = Q \Lambda Q^T \qquad (12.22)$$

where Q is orthogonal with properties

(a) $Q^{-1} = Q^T$

(b) Columns of Q are orthogonal

(c) Columns of Q are normalised eigenvectors

To use SVD factorisation in PCA analysis, the following steps may be followed:

(i) Compute the covariance matrix of the process

(ii) Factorise the covariance matrix into its component matrices

(iii) The singular values of the matrix are used to compute the eigenvalues. They are the square of the singular values: $\lambda_j = \sigma_j^2; \forall_j$

(iv) The columns of the matrix V are read out as the eigenvectors of the covariance matrix

The benefit of SVD is obvious. It provides the eigenvalues, singular values and eigenvectors in one hit.

12.3.4 Applications of PCA

Principal component analysis has found application in a host of object recognition situations of which the most prominent are face recognition, facial expression recognition and noise filtering in images. PCA is also used for channel acquisition in modern telecommunication systems.

12.3.4.1 Face recognition

To use PCA for face recognition, first an average face image is computed from the sum of all images available. The images have the same dimensions (rows and columns). Consider facial images of size 256×256.

(i) From M facial images, form the average image of faces
(ii) For each facial image, convert it from 2D of size N×N to a one-dimensional vector of size $N^2 \times 1$

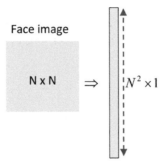

In other words, each 2D face image matrix of dimension N×N is first converted into a vector of dimension $N^2 \times 1$. Therefore, if N is large, this vector can be very long as well. Then, all the long vectors are brought together as a single matrix as illustrated above.

(iii) Form a large matrix of M centred image columns (each column x_i is a facial image for size N×N, e.g., 256×256). Therefore, the Matrix X is of size M×P where P = 256×256 or 64k

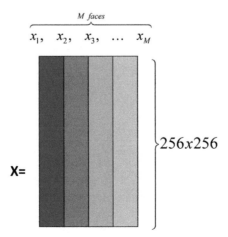

(iv) Compute the covariance matrix Σ

(v) Compute the SVD for the covariance matrix and obtain the eigenvalues (eigenfaces) and eigenvectors

Thus, PCA is used to form a database of features against which a new facial image is tested for recognition.

Reference

[1] M. Turk and, A. Pentland, "Eigenfaces for Recognition", Journal of Cognitive Neuroscience, 3(1), pp. 71–86, 1991.

13

Moment-Generating Functions

13.1 Moments of Random Variables

The statistics and analysis of random variables are normally presented in terms of their moments which include the mean, variance, skew, kurtosis and much more. While the moments of discrete random variables are expressed as summations, the ones for continuous random variables are given as integrals as in Equations (13.1 and 13.2) respectively. We start the discussions by first defining the expectation or expected value of random variables. It is defined as the weighted average the random variable assumes on measurement as expressed for the discrete case as:

$$E[g(X)] = \sum_{x} g(x)p(x) = \sum_{i} g(x_i)p(x_i) \tag{13.1}$$

where $p(x)$ is the probability mass function, $g(X)$ is some function of X, $g(x_i)$ is the value of the function for the ith value of the random variable and $p(x_i)$ is the probability of the ith value of the random variable. For continuous random variable X, the expectation is given as an integral from minus infinity to plus infinity. In other words, the integral is taken over the support of the random variable as

$$E[g(X)] = \int_{-\infty}^{\infty} g(x)f(x)dx \tag{13.2}$$

where $f(x)$ is the probability density function for the random variable. The phrase kth moment of X is used to refer to the mean (first moment, $k = 1$), variance (second moment, $k = 2$), skew (3rd moment, $k = 3$), kurtosis (4th moment, $k = 4$) and so on. This form of terminology is used in this chapter. Let the kth moment of X be given by the expression (13.3):

$$\mu_k = E(X^k) = \begin{cases} \displaystyle\sum_{x \in S} x^k p(x), & \text{if } X \text{ is discrete} \\ \displaystyle\int_{-\infty}^{\infty} x^k f(x)dx, & \text{if } X \text{ is continuous} \end{cases} \tag{13.3}$$

The random variable X takes values within the support (limits) S. The moment when $k = 1$ (the mean) is a useful value for estimating other moments of the random variable. The central moments of the random variable are thus defined by first removing the mean value from the random variable or by centralizing the values to the mean.

13.1.1 Central Moments of Random Variables

Define the kth central moments of a random variable X with the expression

$$\mu_k^0 = E[(X - \mu)^k] = \begin{cases} \displaystyle\sum_{x \in S} (x - \mu)^k p(x), & \text{if } X \text{ is discrete} \\ \displaystyle\int_{-\infty}^{\infty} (x - \mu)^k f(x)dx, & \text{if } X \text{ is continuous} \end{cases}$$

(13.4)

The mean of the random variable is $\mu = \mu_1 = E(X)$. It is also called the first moment of X. All central moments are defined by first subtracting the mean from each element of the random variable and then computing the expectation.

13.1.2 Properties of Moments

(1) The expectation of a constant is the constant. This is stated as E(c)= c, where c is a constant.
(2) Scaling and shift operation. The expectation is:

$$E(aX + b) = aE(X) + b = a\mu + b \qquad (13.5)$$

Both a and b are constants.
(3) The variance of the random variable X is defined by the function

$$\sigma^2 = Var(X) = \mu_2^0 = E(X - \mu)^2$$
$$= E(X^2) - [E(X)]^2 = \mu_2 - \mu_1^2 \qquad (13.6)$$

The standard deviation of the random variable is the square root of the variance, or $\sigma = \sqrt{Var(X)}$.

Proof: By expanding the term in the brackets, we have:

$$E((X - \mu)^2) = E(X^2 - 2\mu X + \mu^2)$$
$$= E(X^2) - 2\mu E(X) + \mu^2$$

$$= E(X^2) - 2\mu^2 + \mu^2$$
$$= E(X^2) - \mu^2$$
$$= E(X^2) - E(X)^2$$

When the random variable takes the discrete values $x_1, x_2, \ldots, x_N$, then the variance is given by the expression

$$Var(X) = E((X - \mu)^2) = \sum_{i=1}^{N} p(x_i)(x_i - \mu)^2 \qquad (13.7)$$

(4) $Var(aX + b) = a^2 Var(X)$

Exercise 1

Given the random variable z in the following table, plot the probability mass function

value z	1	2	3	4	5
pmf $p(z)$	5/10	0	0	0	5/10

Solution

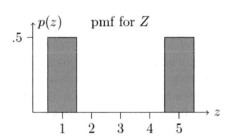

Notice that the sum of the probability mass function is 1.

Exercise 2

Given the mean of the following random variable to be $\mu = 3$, find the

value x	1	2	3	4	5
pmf $p(x)$	1/5	1/5	1/5	1/5	1/5
$(X - \mu)^2$	4	1	0	1	4

(a) variance of the random variable
(b) standard deviation of the random variable
(c) $Var(X) = E((X - \mu)^2) = \frac{4}{5} + \frac{1}{5} + \frac{0}{5} + \frac{1}{5} + \frac{4}{5} = 2.$
(d) the standard deviation is $\sigma = \sqrt{Var(X)} = \sqrt{2}$

In the next section, the expectations of univariate and multivariate random variables are defined in terms of their moment-generating functions. Moment-generating functions are preferred when it is difficult to evaluate the moments due to the complexities of the expressions involved.

13.2 Univariate Moment-Generating Functions

Moment-generating function is defined generally as the expectation of the random variable X when the random variable is the exponential function. For the univariate case, it is defined for random variable as a function of time by the Equation (13.8).

$$M_X(t) = \mathrm{E}\{e^{tx}\} = \begin{cases} \sum_{x \in S} e^{tx} p_X(x); & discrete\ case \\ \int_{x \in S} e^{tx} f(x) dx; & continuous\ case \end{cases} \tag{13.8}$$

The definition is over the support of x or S. Note, $p_X(x)$ is the probability mass function of the random variable. For the continuous case, f(x) is the probability density function. In this chapter, both of them will be used in developing the MGF of both discrete and continuous random variables.

The above definition allows for the computation of distinct n moments of the random variable X, where the expectations for $n = 1$ is the mean, for $n = 2$ it is the variance, $n = 3$ is the skew and $n = 4$ is the Kurtosis of the random variable. Therefore, the nth order expectation can be defined for random variables. Therefore, the following is a general definition of the kth moment of the random variable X. These moments are $E(X), E(X^2), E(X^3), \ldots, E(X^k)$. The MGF are extremely handy in situations when direct computation of these moments proves complex. Therefore, given the MGF each of these moments are given by the kth derivative of the MGF when $t = 0$. In general, the kth moment of a random variable X is defined in terms of the moment-generating function as

$$E(X^k) = \frac{d^k M_X(t)}{dt^k} \bigg|_{t=0} \tag{13.9}$$

This is useful in cases when the derivative with respect to time is possible and can be evaluated for $t = 0$. Therefore, the first derivative gives the mean, the second derivative with respect to time is the variance, the third derivative gives the skew and the fourth derivative is the Kurtosis of the distribution. Note too that one of the essential features of the exponential function e^x is that its derivative is equal to itself. Therefore, consider the exponential function and expand it into power series as follows:

$$e^x = 1 + x + \frac{x^2}{2} + \frac{x^3}{6} + \ldots = \sum_{i=0}^{\infty} \frac{x^i}{i!} \qquad (13.10)$$

The derivative of the function can be shown to be exactly the function itself as:

$$\frac{de^x}{dx} = 0 + 1 + \frac{2x}{2!} + \frac{3x^2}{3!} + \ldots = 1 + x + \frac{x^2}{2!} + \ldots = e^x \qquad (13.11a)$$

Following this process, indeed it is easy to show that the nth derivative of the exponential function is itself:

$$\frac{d^n e^x}{dx^n} = 0 + 1 + \frac{2x}{2!} + \frac{3x^2}{3!} + \frac{4x^3}{4!} + \ldots = 1 + \frac{x}{1!} + \frac{x^2}{2!} + \frac{x^3}{3!} + \ldots = e^x$$

$$(13.11b)$$

Hence, taking the derivative of the exponential function to any power is not necessary to be done as it is itself! Therefore, if we change the exponent of the function by multiplying it with t so that $Y = tX$, it does not really change the nature of the derivative, which is:

$$\frac{de^{tX}}{dt}\bigg|_{t=0} = 0 + X + \frac{2tX^2}{2!} + \frac{3t^2X^3}{3!} + \ldots \bigg|_{t=0}$$

$$= X + tX^2 + \frac{t^2X^3}{2!} + \ldots \bigg|_{t=0} = X + 0 + 0 + \ldots = X$$

Therefore

$$\frac{d^2 e^{tX}}{dt^2}\bigg|_{t=0} = 0 + X^2 + tX^3 + \ldots \big|_{t=0} = X^2 + 0 + \ldots = X^2 \qquad (13.12)$$

Thus, the nth derivative with respect to t is also the variable raised to power n or X^n.

Example 1

Given a discrete random variable X with probability mass function of the random variable X as $p_X(x) = c\left(\frac{1}{3}\right)^X$, where c is a constant, what is the moment-generating function for this distribution? The distribution have values at $x = 0, 1, 2, 3, \ldots, \infty$.

Solution

$$M_X(t) = \mathrm{E}\left\{e^{tx}\right\} = \sum_\Upsilon e^{tx} p_X(x)$$

$$= \sum_\Upsilon e^{tx} \cdot c \left(\frac{1}{3}\right)^x = c \sum_\Upsilon \left(\frac{e^t}{3}\right)^x = c \sum_{x=0}^\infty \left(\frac{e^t}{3}\right)^x$$

Observe that the solution is a geometric series and as such has a known sum. The sum for the geometric series is

$$M_X(t) = c \sum_{x=0}^\infty \left(\frac{e^t}{3}\right)^x = c \sum_{x=0}^\infty (r)^x; \ r = \frac{e^t}{3}$$

$$M_X(t) = c \sum_{x=0}^\infty (r)^x = c \cdot \frac{1}{1-r}; \ r < 1$$

Therefore, we have $M_X(t) = \frac{c}{1-\frac{e^t}{3}}$.

We can therefore find the expectation of the random variable by using derivatives. This is given for the above MGF as

$$E(X) = M_X'(0) = \left.\frac{dM_X(t)}{dt}\right|_{t=0} = \frac{-c}{(1-e^t/3)^2} \cdot (-e^t/3)$$

$$= \left.\frac{c(e^t/3)}{(1-e^t/3)^2}\right|_{t=0} = \frac{c}{3\left(\frac{2}{3}\right)^2} = \frac{3c}{4}$$

13.3 Series Representation of MGF

An exponential function yields to Taylor series expansion readily. Therefore, the MGF can be expanded in Taylor series with expectation taken over the random variable in the series. This is given in this section as an alternative way of expression the MGF.

$$M_X(t) = \mathrm{E}\{e^{tx}\} = \mathrm{E}\left\{\frac{(tx)^0}{0!} + \frac{(tx)^1}{1!} + \frac{(tx)^2}{2!} + \frac{(tx)^3}{3!} + \ldots\right\}$$

$$= E(1) + E(tx) + E\left(\frac{(tx)^2}{2!}\right) + E\left(\frac{(tx)^3}{3!}\right) + \ldots$$

$$= 1 + tE(x) + \frac{t^2}{2!}E(x^2) + \frac{t^3}{3!}E(x^3) + \dots.$$

$$= \sum_{k=0}^{\infty} \frac{t^k}{k!}E(x^k) \tag{13.13}$$

This is so because t is a constant with respect to the expectation.

13.3.1 Properties of Probability Mass Functions

In this section, important properties of probability mass functions are given. Three properties are discussed. Equation (13.1) for both the discrete and continuous MGF depends on the probability mass function. Probabilities are generally positive. Therefore, it is logical and reasonable to assume that pmf will also be positive. In fact, they are generally positive.

(1) The probability mass function of a random variable X is positive for all values x of the random variable within the support. Therefore

$$p_X(x) \geq 0; \quad x \in S$$

When S is finite $p_X(x) > 0$

(2) The sum of the probability mass values for all values x_i of the random variable is unity. That is:

$$\sum_{x_i \in S} p_X(x_i) = 1 \tag{13.14}$$

(3) Given a subset B of the support S of the random variable, the probability of this subset is the sum of the discrete probabilities of all values x_i of the random variable belonging to the subset B. That is

$$Prob(B) = \sum_{x_i \in B} p_X(x_i), \ B \subseteq S \tag{13.15}$$

13.3.2 Properties of Probability Distribution Functions $f(x)$

The probability distribution function $f(x)$ of a random variable X is defined by the expression

$$f_X(t) = Prob(X \leq t)$$

Several properties of the probability distribution function are given in this section.

(1) For negative values of t, the probability distribution function is identically zero. That is

$$\lim_{t \to -\infty} f_X(t) = 0$$

(2) The probability distribution function tends to unity as t tends to positive infinity. That is

$$\lim_{t \to \infty} f_X(t) = 1$$

(3) Since the probability distribution function is an increasingly positive fraction, hence

$$f_X(r) \ge f_X(t), \quad if \; r \ge t$$

This property shows that as the support of the random variable lengthens, the probability distribution function also increases, but is always less than one.

13.4 Moment-Generating Functions of Discrete Random Variables

This section summarizes moment-generating functions for popular distributions. Statistical variables are often modelled with various types of distributions. The discussions in this section are for discrete random variables including Bernoulli, binomial and geometric distributions.

13.4.1 Bernoulli Random Variable

A Bernoulli random variable takes on two values zero (0) and one (1) with probabilities, p for 'success' and the second $1 - p$ for 'failure'. Therefore, its moment generation function is defined as

$$M_X(t) = M_X(e^{tX}) = \sum_{x=0}^{1} e^{tx} p(x) = e^{0.t}(1 - p) + e^{1.t}p \tag{13.16}$$

Therefore

$$M_X(e^{tX}) = (1 - p) + e^t p \tag{13.17}$$

We have shown in Section 13.1 that the nth derivative of an exponential function is also the exponential function. For the Bernoulli distribution, the nth moment as given by the moment-generating function. It can therefore be shown very easily to be

$$\chi_m = \left.\frac{d^m M(t)}{dt^n}\right|_{t=0} = pe^t|_{t=0} = p$$

Thus, all the moments are equal as the probability of a success.

Consider the definition of the moment-generating function of the random variable X:

$$M(t) = E(e^{tX}) = \sum_{x \in S} e^{tx} f(x) \tag{13.18}$$

S is the support of the random variable. If the support is given by the sequence $\{a_1, a_2, a_3, \ldots\}$, then we can define the moment-generating function as a sum as follows:

$$M(t) = e^{ta_1} f(a_1) + e^{ta_2} f(a_2) + e^{ta_3} f(a_3) + \ldots. \tag{13.19}$$

where the mass function are the coefficients or the probabilities when the random variable takes the values within the support.

$$f(a_j) = P(X = a_j)$$

Two random variables with identical moment-generating functions also have the same probability mass function.

13.4.2 Binomial Random Variables

Binomial random variable X represents a process which repeats the Bernoulli process for a number of times, for example, the number of successes in n trials of a Bernoulli process. The probability of k successes in n trials, where p represents a success given by the expression

$$P_X(k) = \text{Prob}(X = k) = C_k^n p^k (1-p)^{n-k} \tag{13.20}$$

The binomial distribution is represented as b(n, p). The expectation of the process is given by the expression

$$E[X] = \sum_{k=0}^{n} k C_k^n p^k (1-p)^{n-k} = \sum_{k=0}^{n} \frac{n!}{(k-1)!(n-k)!} p^k (1-p)^{n-k}$$

$$\tag{13.21}$$

The variance of the process is also given as

$$E[X^2] = \sum_{k=0}^{n} k^2 C_k^n p^k (1-p)^{n-k} = \sum_{k=0}^{n} \frac{n!k}{(k-1)!(n-k)!} p^k (1-p)^{n-k}$$

(13.22)

Both expressions contain a number of terms to be computed. This complexity can be avoided by using the moment-generating function of the binomial distribution. The moment-generating function for a binomial random variable is defined by the expression

$$M(t) = [(1-p) + pe^t]^n$$

(13.23)

This expression may be derived through the following expression

$$M(t) = E(e^{tX}) = \sum_{x=0}^{n} e^{tx} \left[\binom{n}{x} p^x (1-p)^{n-x} \right]$$

(13.24)

By re-arranging the terms, we have

$$M(t) = \sum_{x=0}^{n} \binom{n}{x} (pe^t)^x (1-p)^{n-x}$$

$$= \sum_{x=0}^{n} \binom{n}{x} a^x b^{n-x}$$

$$= (a+b)^n$$

$$= (pe^t + (1-p))^n$$

(13.25)

Exercise: Calculate the first and second moments of a binomial random variable.

Example 1

If a random variable X has the following moment-generating function:

$$M(t) = \left(\frac{3}{4} + \frac{e^t}{4} \right)^{20}$$

for all t, what is the p.m.f. of $X (-\infty < t < \infty)$? The moment-generating function given looks like that for a binomial random variable. This means therefore that the random variable X is binomial and $n = 20$ while $p = 1/4$. The probability mass function for X is therefore

$$f(x) = \binom{20}{x} \left(\frac{1}{4} \right)^x \left(\frac{3}{4} \right)^{20-x}$$

For $x = 0, 1, \ldots, 20$.

Example

If a random variable X has the following moment-generating function:

$$M(t) = \frac{1}{10}e^t + \frac{2}{10}e^{2t} + \frac{3}{10}e^{3t} + \frac{4}{10}e^{4t}$$

for all t, what is the p.m.f. of X? The probability mass function is therefore

$$f(x) = \begin{cases} \dfrac{1}{10}, & if \ x = 1 \\[2mm] \dfrac{2}{10}, & if \ x = 2 \\[2mm] \dfrac{3}{10}, & if \ x = 3 \\[2mm] \dfrac{4}{10}, & if \ x = 4 \end{cases}$$

13.4.3 Geometric Random Variables

In this section, we derive the expression for the moment-generating function for a geometric random variable. By definition

$$M(t) = E[e^{tX}] = \sum_x e^{tx}(q^{x-1}p) = e^t p + e^{2t}qp + e^{3t}q^2 p + e^{4t}q^3 p + \cdots$$

$$(13.26)$$

We will re-write this as a geometric series and then sum the geometric series using standard summation technique to a geometric series. This becomes

$$M(t) = e^t p(1 + e^t q + e^{2t}q^2 + e^{3t}q^3 + \cdots)$$

$$= e^t p \sum_n (e^t q)^n = e^t p \left(\frac{1}{1 - e^t q}\right)$$

$$= \frac{pe^t}{1 - qe^t} \qquad (13.27)$$

Therefore, we can generate the moments a lot more easily by taking the derivative of M(t) with respect to time and evaluating it when $t = 0$ as is the norm.

Exercise: Given the expression for the MGF for a geometric random variable as in this section, what are the first and second moments of the random variable?

13.4.4 Poisson Random Variable

Poisson processes model the occurrence of events within a time duration. For example, the arrival of passengers at a train station is a Poisson process. Another example is the arrival rates of traffic at a point whether this is data or vehicular traffic does not matter. The arrival rates at two different time durations are also independent of each other. In the limit, the probability of the process occurring depends on the arrival rate with a short time interval. Although the occurrence of the process is not predictable,

$$P(t \rightarrow t + \Delta t) = \lambda.\Delta t$$
$$\lim_{\Delta t \rightarrow 0} \tag{13.28}$$

For example, the number of access of a web page between 4pm and 5pm is a Poisson process. To model the process well, the time interval is divided into n discrete time intervals which allow for estimation of the number of arrivals for each of the time interval.

Note that λ is the arrival rate of the process.

$$M(t) = E(e^{tY}) = \sum_{y=0}^{\infty} e^{ty} \left[\frac{e^{-\lambda}\lambda^{y}}{y!} \right] = \sum_{y=0}^{\infty} \frac{e^{-\lambda}(\lambda e^{t})^{y}}{y!}$$

$$= e^{-\lambda} \sum_{y=0}^{\infty} \frac{(\lambda e^{t})^{y}}{y!} = e^{-\lambda}e^{\lambda e^{t}} = e^{\lambda(e^{t}-1)} \tag{13.29}$$

Telephone traffic theory depends to a great extent on this modelling technique and has had wide applications. Hence, an easier means of determining the mean arrival rate and variance using the moment-generating function is significantly useful.

13.5 Moment-Generating Functions of Continuous Random Variables

This section summarizes moment-generating functions for popular continuous distributions. The discussions include normal, exponential and Gamma distributions.

13.5.1 Exponential Distributions

An exponential event describes the waiting time between occurrence of Poisson events. For example, a hospital discovers that on average records one birth per week. If T is the interval between births, it has an exponential distribution. A random variable T is said to have an exponential distribution if it has a probability density function defined by the expression

$$f_T(t) = \begin{cases} \lambda e^{-\lambda t} & if \ t > 0 \\ 0 & if \ t \leq 0 \end{cases} \tag{13.30}$$

$\lambda > 0$ is the rate of the distribution. In the hospital example given at the beginning of this section, the pdf is given by the expression

$$f_T(t) = \begin{cases} \dfrac{1}{7} e^{-\frac{1}{7}t} & t > 0 \\ 0 & \text{otherwise} \end{cases} \tag{13.31}$$

$$M(t) = E(e^{tY}) = \int_0^\infty e^{ty}(\lambda e^{-\lambda y})dy = \lambda \int_0^\infty e^{-y(\lambda - t)}dy$$

$$= \lambda \int_0^\infty e^{-y(\lambda(1-t/\lambda))}dy = \lambda \int_0^\infty e^{-y\lambda^*}dy \quad \text{where } \lambda^* = \lambda - t \tag{13.32}$$

The moment-generating function is therefore given as equation (13.33):

$$M(t) = \left(\frac{-\lambda}{\lambda^*}\right)e^{-y\lambda^*}\Big|_0^\infty = \left(\frac{-\lambda}{\lambda^*}\right)(0-1) = \left(\frac{\lambda}{\lambda^*}\right) = \frac{\lambda}{\lambda - t} \tag{13.33}$$

13.5.2 Normal Distribution

Normal distributions are arguably the most widely used statistical distributions in various applications and in almost all fields of endeavour. A normal distribution is also called a Gaussian distribution and used widely in modelling additive white noise in mobile communication channels. A normal distribution of a random variable X having variance σ^2 and mean μ is represented with the probability density function

$$p_X(x) = \frac{1}{\sqrt{2\pi\sigma^2}} \exp\left\{-\frac{1}{2}\frac{(x-\mu)^2}{\sigma^2}\right\} \tag{13.34}$$

By following the definition for moment-generating functions, the moment-generating function for a normal distribution is given by the following expression

$$M(t) = E(e^{tX}) = \int_{-\infty}^{\infty} e^{tx} \left(\frac{1}{\sqrt{2\pi\sigma^2}} \exp\left\{ -\frac{1}{2} \frac{(x-\mu)^2}{\sigma^2} \right\} \right) dx$$

$$= \frac{1}{\sqrt{2\pi\sigma^2}} \int_{-\infty}^{\infty} \exp\left\{ -\frac{x^2}{2\sigma^2} + \frac{x\mu}{\sigma^2} - \frac{\mu^2}{2\sigma^2} + tx \right\} dx \qquad (13.35)$$

By using the expression $(\mu + t\sigma^2)^2 = \mu^2 + 2\mu t\sigma^2 + (t\sigma^2)^2$, we can further simplify the above integral to equation (13.36):

$$M(t) = \frac{1}{\sqrt{2\pi\sigma^2}} \int_{-\infty}^{\infty} \exp\left\{ -\frac{x^2}{2\sigma^2} + \frac{x(\mu + t\sigma^2)}{\sigma^2} - \frac{\mu^2}{2\sigma^2} \right.$$

$$\left. -\frac{2\mu t\sigma^2 + (t\sigma^2)^2}{2\sigma^2} + \frac{2\mu t\sigma^2 + (t\sigma^2)^2}{2\sigma^2} \right\} dx \qquad (13.36)$$

When we also isolate terms under the integral which are not functions of the variable x and take them outside the integral, this results in the simplification in equation (13.37):

$$M(t) = \exp\left\{ \frac{2\mu t\sigma^2 + (t\sigma^2)^2}{2\sigma^2} \right\} \int_{-\infty}^{\infty} \frac{1}{\sqrt{2\pi\sigma^2}}$$

$$\exp\left\{ -\frac{1}{2} \left(\frac{[x - (\mu + t\sigma^2)]^2}{\sigma^2} \right) \right\} dx \qquad (13.37)$$

Since we integrate over the probability density of a random variable from infinity to infinity, the integral term evaluates to 1. In other words

$$\int_{-\infty}^{\infty} \frac{1}{\sqrt{2\pi\sigma^2}} \exp\left\{ -\frac{1}{2} \left(\frac{[x - (\mu + t\sigma^2)]^2}{\sigma^2} \right) \right\} dx = 1 \qquad (13.38)$$

The expression for the probability density function of the random variable inside the integral is often written as $X \sim N(\mu + t\sigma^2, \sigma^2)$. The standard normal distribution has zero mean and unit variance $\mu = 0, \sigma^2 = 1$.

From the simplifications, the moment-generating function for a random variable X having normal distribution simplifies to the expression

$$M(t) = \exp\left\{ \frac{2\mu t\sigma^2 + (t\sigma^2)^2}{2\sigma^2} \right\} = \exp\left\{ \mu t + \frac{t^2\sigma^2}{2} \right\} \qquad (13.39)$$

This expression is a lot easier to use in the computation of the statistics of a normal distributed random variable X. The expression is not a function of x but rather a function of t. It is a lot easier to use in computing the statistics of the random variable.

13.5.3 Gamma Distribution

In this section, we examine the gamma distribution of a random variable X. The distribution of a random variable X is a gamma function if the probability density function of X given two variables $(\alpha, \beta) > 0$ is

$$p(x|\alpha, \beta) = \frac{\beta^\alpha}{\Gamma(\alpha)} x^{\alpha-1} e^{-\beta x} \qquad (13.40)$$

where $x \geq 0$ is continuous. The gamma function $\Gamma(\alpha)$ is a generalization of the factorial function for the continuous case. It is normally considered as the continuous analogue of the factorial function whereby it is defined as equation (13.41):

$$\Gamma(\alpha) = \int_0^\infty t^{\alpha-1} e^{-t} dt \qquad (13.41)$$

By integration, it can be shown that $\Gamma(1) = 1$ and $\Gamma(2) = 1\Gamma(1) = 1$. Similarly,

$$\Gamma(3) = 2\Gamma(2) = 2$$
$$\Gamma(4) = 3\Gamma(3) = 3 \times 2 = 6$$

Therefore, in general, $\Gamma(\alpha) = (\alpha - 1)!$ Indeed, also $\Gamma(\alpha + 1) = \alpha\Gamma(\alpha)$.

The value of the random variable x may be considered as the mean measure of a countable variable. This is reasonable particularly when the mean is a non-integer but positive. Gamma functions have been used as the prior of Bayes inferences and also as prior for a precision parameter. For this case, we consider precision as the inverse of the variance of the parameter.

The moment-generating function for a gamma distribution of a random variable X is given by the expression (13.42):

$$M(t) = E(e^{tX}) = \int_0^\infty e^{tx} \left(\frac{1}{\Gamma(\alpha)\beta^\alpha} x^{\alpha-1} e^{-x/\beta} \right) dx$$

$$= \frac{1}{\Gamma(\alpha)\beta^\alpha} \int_0^\infty x^{\alpha-1} e^{-x\left(\frac{1-\beta t}{\beta}\right)} dx \qquad (13.42)$$

By using the variable $\beta' = \frac{\beta}{1-\beta t}$, we can show that the integral

$$M(t) = \frac{1}{\Gamma(\alpha)\beta^\alpha} \int_0^\infty x^{\alpha-1} e^{-x/\beta'} dx \qquad (13.43a)$$

The term under the integral in equation (a) is actually a gamma function without the normalizing constant as in the definition of a gamma function. That is, by inserting the normalizing constant we have

$$\int_0^\infty x^{\alpha-1} e^{-x/\beta'} dx = \Gamma(\alpha)(\beta')^\alpha \qquad (13.43b)$$

Therefore

$$M(t) = \frac{1}{\Gamma(\alpha)\beta^\alpha} \Gamma(\alpha)(\beta')^\alpha = (1 - \beta t)^{-\alpha} \qquad (13.44)$$

13.6 Properties of Moment-Generating Functions

From the derivations of the moment-generating functions obtained in the previous sections, the following three properties can be identified.

$$(1) \quad M_X(0) = 1$$

At $t = 0$, the values of the moment-generating functions are all equal to one. The reader should check this for all the MGF given so far.

$$(2) \quad M_X^{(k)}(0) = \left. \frac{d^k}{dt^k} M_X(t) \right|_{t=0} = \mu_k$$

The kth moment is obtained by evaluating the kth derivative of the MGF to zero.

$$(3) \quad M_X(t) = 1 + \mu_1 t + \frac{\mu_2}{2!} t^2 + \frac{\mu_3}{3!} t^3 + \cdots + \frac{\mu_k}{k!} t^k + \cdots$$

where $\mu_k = E(X^k)$. Moment-generating functions can be expanded as power series as shown.

13.7 Multivariate Moment-Generating Functions

Analysis of moment-generating functions so far has been on single-variate distributions. This can be extended to multivariate random variables. Many processes occur together resulting to several random variables occurring at the same time. Multiple random variables may be independent or dependent. Consider the set of random variables $X_1, \ldots, X_k$. Assume they are independent. The expectation of the random variables is therefore given as

$$
\begin{aligned}
M_{X_1+X_2+\cdots+X_K}(t) &= E(X_1 + X_2 + \cdots + X_K) \\
&= E(e^{(X_1+X_2+\cdots+X_K)t}) \\
&= E(e^{X_1 t} e^{X_2 t} e^{X_3 t} \cdots e^{X_K t}) \\
&= E(e^{X_1 t}) E(e^{X_2 t}) \cdots E(e^{X_K t}) \\
&= M_{X_1}(t) M_{X_2}(t) \cdots M_{X_K}(t) \qquad (13.45)
\end{aligned}
$$

This result is obtained by considering that the random variables are independent. Therefore

$$
E(X_1 + X_2 + \cdots + X_K) = \prod_{k=1}^{K} M_{X_k}(t) \qquad (13.46)
$$

13.7.1 The Law of Large Numbers

The law of large numbers applies to multivariate distributions consisting of an independent trials process $Y_k = X_1 + X_2 + \cdots + X_k$. Let each unit of the trial process have the unit mean and variance $\mu = E(X_j)$ and $\sigma^2 = E(X_j^2)$, the law of large numbers states that as n tends to infinity the value of this probability tends to zero

$$
P\left(\left|\frac{Y_k}{k} - \mu\right| \geq \varepsilon\right) \to 0
$$

However, also as n tends to infinity, the value tends to one.

$$
P\left(\left|\frac{Y_k}{k} - \mu\right| < \varepsilon\right) \to 1
$$

The proof of this the law of large numbers comes from the following simple consideration. Since all the random variables in Y are independent and have the same distributions, they also have the same mean and variance, it means that the variance of Y is the sum of all the variances of members of the random variable set. That is, the variance of Y is $k\sigma^2$ or

$$
Var(Y) = k\sigma^2
$$

Also the

$$Var\left(\frac{Y}{k}\right) = \frac{\sigma^2}{k}$$

Note that the ratio Y/k is the average of unit outcomes of the process. Chebyshev's inequality for any value of $\varepsilon > 0$ can be used to show that

$$P\left(\left|\frac{Y_k}{k} - \mu\right| \geq \varepsilon\right) < \frac{\sigma^2}{k\varepsilon^2}$$

Thus, for fixed value of ε, the two expressions for the law of large numbers are true.

13.8 Applications of MGF

One area where MGF can be applied is in the computation of market values of depreciated products after a length of time.

The Birnbaum–Saunders distribution is a fatigue life distribution that was derived from a model assuming that failure is due to the development and growth of a dominant crack. This distribution has been shown to be applicable not only for fatigue analysis but also in other areas of engineering science. Because of its increasing use, it would be desirable to obtain expressions for the expected value of different powers of this distribution.

14

Characteristic Functions

14.1 Characteristic Functions

Transforming problems from one domain to the other (e.g., time to frequency) often provides a new perspective for looking at the characteristics of a signal. Characteristic function is a typical example whereby the probability density function (pdf) of a random variable is transformed into a new space and provides an avenue for studying the pdf better. The characteristic function of a random variable in probabilistic language is the Fourier transform of the probability distribution function of the random variable. Characteristic functions are generalisations of moment-generating function by extending it to the complex number space.

The characteristic function of a random variable X is unique. Therefore, if by creating the characteristic functions of two random variables X and Y and they are the same, then the processes X and Y represent the same process.

Characteristic functions are very close in definition to that of moment generating functions with one unique difference. The exponential function is complex and therefore provides a mixture of a complex exponential function and a real probability density function being multiplied and integrated together.

In general, the characteristic function of a random variable X is defined as:

$$\varphi_X(t) = E(e^{jxt}) = \int_{-\infty}^{\infty} e^{jxt} f(x)\, dx \qquad (14.1)$$

where $j = \sqrt{-1}$. Due to the mixture of real and complex terms in the integral, some form of manipulation is required. In the rest of this section, examples will be given on how this is done.

239

14.1.1 Properties of Characteristic Functions

Consider the random variables $X_1, X_2, \ldots, X_K$. If they are independent random variables, then

(1) The characteristic function of the sum of independent random variables is the product of their characteristic functions, where

$$\varphi_{X_1, X_2, \ldots, X_K}(t) = \prod_{k=1}^{K} \varphi_{X_k} \tag{14.2}$$

(2) $\varphi_{aX}(t) = \varphi_X(at)$
(3) The characteristic function is continuous for continuous random variables

This result comes directly from the definition of the characteristic function of the random variable X.

Proof: Let

$$\varphi_{aX}(t) = E(e^{jaxt}) = \int_{-\infty}^{\infty} e^{jaxt} f_X(x) \, dx = \varphi_X(at) \tag{14.3}$$

This is because the variable t is now multiplied by a in the exponential.

(4) $\varphi_X(0) = 1$ and $|\varphi_X(t)| \le 1$ for all t.
(5) For all t such that $t_t < t_2 < \cdots < t_n$, the matrix formed by the elements of the characteristic function $A = (a_{ij})_{1 \le i,j \le n}$ so that $a_{jk} = \varphi_X(t_j - t_k)$ is Hermittian and also positive definite. This means that $A^* = A$, for $\zeta^T A \bar{\zeta} \ge 0$, for any $\zeta \in \mathbb{C}^n$.

14.2 Characteristic Functions of Discrete Single Random Variables

In this section, we derive the characteristic functions of discrete random variables X. The most fundamental discrete random variables are covered with examples.

14.2.1 Characteristic Function of a Poisson Random Variable

For a Poisson random variable, the probability mass function is

$$p(x_k) = p(X = k) = e^{-\lambda} \frac{\lambda^k}{k!}, \quad k = 0, 1, 2, \ldots, \infty. \tag{14.4}$$

Therefore, its characteristic function is defined by the expression

$$\varphi_X(t) = \sum_{k=1}^{K} e^{jkt} p(x_k).$$ (14.5)

Therefore, the characteristic function for a Poisson random variable is

$$\varphi_X(t) = \sum_{k=0}^{\infty} e^{jkt} e^{-\lambda} \frac{\lambda^k}{k!}$$

$$= e^{-\lambda} \sum_{k=0}^{\infty} \frac{(\lambda e^{jt})^k}{k!} = e^{-\lambda} \sum_{k=0}^{\infty} \frac{x^k}{k!}$$

$$= e^{-\lambda} e^{\lambda jt} = e^{\lambda(e^{jt}-1)}$$ (14.6)

This result comes from the fact that the sum $\sum_{k=0}^{\infty} \frac{(\lambda e^{jt})^k}{k!}$ is a Taylor series with variable $x = (\lambda e^{jt})$. To find the moments of the random variable, we expand the series and take derivatives and evaluate the derivatives at $t = 0$. Since

$$\varphi(e^{jtX}) = E\left(\sum_{k=0}^{\infty} \frac{(\lambda e^{jt})^k}{k!}\right)$$ (14.7)

14.2.2 Characteristic Function of Binomial Random Variable

If a random process has binomial distribution, the characteristic function is given by a binomial function:

$$\varphi_X(t) = \sum_{k=0}^{n} e^{jkt} \binom{n}{k} p^k q^{n-k}$$

$$= \sum_{k=0}^{n} \binom{n}{k} (pe^{jt})^k q^{n-k}$$

$$= (pe^{jt} + q)^n$$ (14.8)

The probabilities p and q are estimated from data directly. For example, a binomial process when p represents the probability of success and q the probability of a failure is modelled by Equation (14.8).

14.2.3 Characteristic Functions of Continuous Random Variables

We start with an obvious example, the characteristic function of a normal random variable with probability density function:

$$f_X(x) = \frac{1}{\sqrt{2\pi}} e^{-x/2} \tag{14.9}$$

Therefore, the characteristic function is given by the expression

$$\varphi_X(t) = E(e^{jxt}) = \int_{-\infty}^{\infty} e^{jxt} f_X(x)\, dx$$

$$= \int_{-\infty}^{\infty} e^{jtx} \frac{1}{\sqrt{2\pi}} e^{-x/2} dx \tag{14.10}$$

To help simplify the integration, manipulation of the terms in the integral is required. This is done by pre-multiplying it with an identity exponent as shown

$$\varphi_X(t) = \int_{-\infty}^{\infty} \frac{e^{-t^2/2} e^{t^2/2}}{\sqrt{2\pi}} e^{jtx} e^{-x^2/2} dx$$

$$= \frac{e^{-t^2/2}}{\sqrt{2\pi}} \int_{-\infty}^{\infty} e^{-(x-jt)^2/2} dx \tag{14.11}$$

This integral requires a change of variable. This is done by letting $z = x - jt \Rightarrow dz = dx$ and then integrating over the new limits, which are $z_{upper} = \infty - jt$; $z_{lower} = -\infty - jt$. This becomes

$$\varphi_X(t) = \frac{e^{-t^2/2}}{\sqrt{2\pi}} \int_{-\infty-jt}^{\infty-jt} e^{-z^2/2} dz \tag{14.12}$$

We will integrate this function over a contour using residues. It is known that the integral over a closed contour yields the result

$$\oint f(z)dz = 2\pi j \sum \text{residues} = 0 \tag{14.13}$$

Thus, the residues are zero if the contour is closed.

15

Probability-Generating Functions

15.1 Probability-Generating Functions

No single mathematical tool is adequate for most of the data processing requirements in many areas of study. Hence, having a range of tools to deploy in solving a problem can be very useful. When one tool fails or when those tools become difficult to use, other tools may be used by the processor. This is one of the many reasons for analysis of random variables using probability-generating functions.

This chapter explores another approach leading to computation of expectations. It provides a fourth method for computing moments in addition to the direct method, moment-generating function (MGF) and characteristic functions. In the next section, we consider the discrete case for the probability-generating functions. As in the analysis for MGF and characteristic functions, popular distributions are considered to illustrate the approach. The phrase "probability-generating function" is a misnomer. It is a series created from probability distribution based on integers. The function itself does not and is not used to create the probabilities when values are substituted for s^k in Equation (15.1). Therein is the misnomer.

15.2 Discrete Probability-Generating Functions

Consider a random variable X, with values within the real number set. The probability-generating function for the random variable is defined with the expression

$$G_X(s) = E(s^X) = \sum_k s^k P(X = x_k) \tag{15.1}$$

For ease of analysis, we rewrite this expression with the notation

$$G_X(s) = E(s^X) = \sum_k s^k P(X = k) \tag{15.2}$$

243

where k is a positive number, usually an integer. In other words, we are interested in computing the expectations of the random variable X through these definitions. The definition makes no assumptions about the support of X. The support may be large or infinite. If the support is finite as many random variables are, then X takes a finite set of values within the support as defined by Equations (15.1 and 15.2).

Let us explore further the case for which X takes a finite set of values $x_0, x_1, x_2, \ldots, x_N$, the probability-generating function (PGF) is then a polynomial in s with $(N+1)$ terms. The polynomial yields to solutions using the various methods for solving polynomial functions.

$$G_X(s) = E(s^X) = \sum_{k=0}^{N} s^k P(X = k)$$

$$= P(X = 0) + P(X = 0)s + \cdots + P(X = N)s^N \qquad (15.3)$$

When the random variable X takes a large countable set of values $x_0, x_1, x_2, \ldots, x_N, \ldots$ we obtain a series which converges to some value, which is always less than unity, provided the magnitude of s is less than or equal to one ($|s| \leq 1$ when $s \in [-1, 1]$.

$$G_X(s) = E(s^X) = \sum_{k=0} s^k P(X = k)$$

$$= P(X = 0) + P(X = 0)s + \cdots + P(X = N)s^N + \cdots \qquad (15.4)$$

This is because the following relationship holds

$$G_X(s) = \sum_{k \geq 0} s^k P(X = k) \leq \sum_{k \geq 0} \left| s^k \right| P(X = k) \leq \sum_{k \geq 0} P(X = k) = 1$$

$$(15.5)$$

To show how probability-generating functions may be used we illustrate the method by deriving the PGF for some of the popular distributions. We start with the simplest of them, the Bernoulli random variable.

15.2.1 Properties of PGF

(1) Moments of random variable X

If the radius of convergence R of the probability-generating function for a random variable X is greater than 1 ($R > 1$), moments of a random variable may be computed from the probability-generating functions by taking appropriate derivative and evaluating it at $s = 1$.

(a) The mean or the first moment of the random variable is

$$E(s^X) = G'_X(s) = \frac{d}{ds}\left[\sum_{k=0} s^k P(X = k)\right] = \sum_{k\geq 1} k s^{k-1} P(X = k)$$

$$(15.6)$$

Observe the summation of the derivative is taken from k equal to or more than one. This is due to the derivative of the first term at $k = 0$ (a constant) being zero.

The mean or expectation of the random variable is

$$E(X) = G'_X(1) = \sum_{k\geq 1} k \times s^{k-1} P(X = k) = \sum_{k\geq 1} k \times P(X = k)$$

$$(15.7)$$

(b) The kth moment of the random variable is

$$G_X^{(k)}(1) = \lim_{s\to 1} G_X^{(k)}(s) \qquad (15.8)$$

The kth moment can be derived from the expression

$$E(X(X - 1)\cdots(X - k + 1)) = G_X^{(k)}(1) = \lim_{s\to 1} G_X^{(k)}(s)$$

$$(15.9)$$

which means that we need to first derive the expectation $E(X(X - 1))$ before and the first moment before arriving at the variance of the random variable. Thus, once more we have another approach for computing the moments of random variables.

Example: Consider the following polynomial which models the probability-generating function of a random variable. $G_X(s) = p_0 + sp_1 + s^2 p_2 + s^3 p_3 + s^4 p_4$

(i) Find the first, second, third and fourth moments of the random variable

Solution
The first moment of the random variable is $\frac{d}{ds}G_X(s) = p_1 + 2sp_2 + 3s^2 p_3 + 4s^3 p_4$. When evaluated at $s = 1$, this becomes $E(X) = G_X(1) = p_1 + 2p_2 + 3p_3 + 4p_4$.

The second moment of the random variable is

$$\frac{d^2}{ds^2}G_X(s)\bigg|_{s=1} = 2p_2 + 3 \times 2sp_3 + 4 \times 3s^2 p_4 = 2p_2 + 6p_3 + 12p_4$$

The third moment-generating function of the random variable is

$$\frac{d^3}{ds^3}G_X(s)\bigg|_{s=1} = 3 \times 2p_3 + 4 \times 3 \times 2sp_4 = 6p_3 + 24p_4$$

The fourth moment-generating function becomes

$$\frac{d^4}{ds^4}G_X(s)\bigg|_{s=1} = 4 \times 3 \times 2p_4 = 24p_4$$

15.2.2 Probability-Generating Function of Bernoulli Random Variable

Consider an unbiased coin tossing random variable with probabilities p and q such that $p + q = 1$. The probability-generating function is given by the expression

$$G_X(s) = \sum_{k=0}^{1} s^k P(X = k)$$

$$= P(X = 0) + sP(X = 1) = q + sp \qquad (15.10)$$

Example 1: If for a Bernoulli random variable X, we have the probabilities $p = 1/2$ and $q = 1/2$, write an expression for the Bernoulli probability-generating function.

Solution:
From Equation (15.6), we are given that $G_X(s) = \sum_{k=0}^{1} s^k P(X = k) = \frac{1}{2} + \frac{1}{2}s$.

Exercise 1: Normally for an unbiased coin, s is limited to $s = 1$. Suppose s takes real values less than 1, sketch the probability-generating function for the Bernoulli distribution as a function of s between 0 and 1.

15.2.3 Probability-Generating Function for Binomial Random Variable

The binomial random variable X occurs in a discrete experiment where n experiments are undertaken with k successes with probability p. The remaining outcomes $(n - k)$ result to failures with probability q such that $p + q = 1$.

Consider the probability mass function when k successes occur. This is given by the expression

$$P(X = k) = \binom{n}{k} p^k q^{n-k}, \quad k = 0, 1, 2, \ldots, n \tag{15.11}$$

The probability-generating function is given from the definition of PGF as

$$G_X(s) = \sum_{k \geq 0} s^k P(X = k) = \sum_{k \geq 0} s^k \binom{n}{k} p^k q^{n-k}$$

$$= \sum_{k=0}^{n} \binom{n}{k} (sp)^k q^{n-k} = (q + sp)^n, \quad s \in \mathbb{R} \tag{15.12}$$

The expectation for the binomial random variable using its PGF is

$$E(X) = G'_X(1) = \frac{d}{ds}(q + sp)^n \Big|_{s=1} = np(q + sp)^{n-1}\Big|_{s=1}$$

$$= np(q + p)^{n-1} \tag{15.13}$$

Since $q + p = 1$, the expectation simplifies to

$$E(X) = G'_X(1) = np \tag{15.14}$$

15.2.4 Probability-Generating Function for Poisson Random Variable

Given the Poisson rate λ, the PGF is given by the expression

$$P(X = k) = e^{-\lambda} \frac{\lambda^k}{k!}, \quad k = 0, 1, 2, \ldots \tag{15.15}$$

This leads to the expression for the probability-generating function in Equation (15.16):

$$G_X(s) = \sum_{k \geq 0} s^k P(X = k) = e^{-\lambda} \sum_{k \geq 0} s^k \frac{\lambda^k}{k!}, \quad k = 0, 1, 2, \ldots$$

$$= e^{-\lambda} \sum_{k=0}^{\infty} \frac{(s\lambda)^k}{k!} = e^{-\lambda} e^{s\lambda}$$

$$= e^{(s-1)\lambda} \tag{15.16}$$

The result from the power series expansion is the summation which is $e^{s\lambda}$. We will now also derive the expectation of the Poisson random variable using the PGF. The expectation is the first derivative of the PGF evaluated at $s = 1$. This is shown in Equation (15.17).

$$E(X) = G'_X(1) = \frac{d}{ds} e^{(s-1)\lambda} \Big|_{s=1} = \lambda e^{(s-1)\lambda}|_{s=1} = \lambda \qquad (15.17)$$

The second moment may be derived equally through the interim expression. The interim expectation is that of a product of the random variable and the random variable minus 1. That is given in Equation (15.18) and also evaluated at $s = 1$.

$$E(X(X-1)) = G''_X(1) = \frac{d^2}{ds^2} e^{(s-1)\lambda} \Big|_{s=1} = \lambda^2 e^{(s-1)\lambda}|_{s=1} = \lambda^2$$
$$(15.18)$$

Given that the variance of a random variable is given by the expression

$$Var(X) = E(X^2) - (E(X))^2 \qquad (15.19)$$

From our analysis in Equation (15.19), we have

$$E(X(X-1)) = E(X^2) - E(X) = \lambda^2$$

By using Equations (15.19) and (15.18), the variance of the Poisson random variable is

$$E(X^2) = E(X(X-1)) + E(X) = \lambda^2 + \lambda \qquad (15.20)$$

Both the variance and mean of the random variable are defined by the arrival rate of the Poisson random variable. Therefore, knowing the arrival rate means we know a lot about the process right from the beginning.

15.2.5 Probability-Generating Functions of Geometric Random Variables

A geometric random variable and indeed **geometric distribution** is a special case of the negative binomial random variable (distribution). A negative binomial distribution is used to analyse the number of trials before one success is achieved. Therefore, a geometric distribution is a negative binomial distribution with $r = 1$ success.

As an example, how many times do we have to toss an unbiased coin before we record one head, where a head is taken as success? Another good example of a geometric random variable is how many times does an IVF doctor undertake an IVF implantation on a female patient before she becomes pregnant. The success is her being pregnant. Suppose the IVF doctor tries k times before a pregnancy results, we can represent the probability distribution for the random process X as

$$g(x; p) = q^{k-1}p \tag{15.21}$$

The geometric random variable possesses two parameters p and q which are the probabilities of one occurrence of each type of outcome (success or failure). The probability mass function is defined by the expression

$$P(X = k) = q^{k-1}p, \quad k = 0, 1, 2, \ldots; \quad 0 < p, q < 1 \tag{15.22}$$

Like all other distributions, it also has a unique probability-generating function given in Equation (15.23)

$$G_X(s) = \sum_{k=1}^{\infty} s^k q^{k-1} p = sp \sum_{k=1}^{\infty} s^{k-1} q^{k-1} p$$

$$= sp \sum_{k=1}^{\infty} (sq)^{k-1} = \frac{sp}{1 - sq}, \quad |s| < \frac{1}{q} \tag{15.23}$$

The term under the summation is a geometric series and lead to the result. The result is valid provided the magnitude of q is less than 1/q.

The expectation using the PGF is derived the same way as for the other distributions discussed already and is given by the expression

$$E(X) = G'_X(1) = \frac{d}{ds} \frac{sp}{1 - sq} \bigg|_{s=1}$$

$$= \frac{p}{(1 - sq)^2} \bigg|_{s=1} = \frac{1}{p} \tag{15.24}$$

The mean of the distribution is the inverse of the probability of a success p. The variance of the distribution is also obtained by first deriving the expectation:

$$E(X(X - 1)) = G''_X(1) = \frac{d^2}{ds^2} \frac{sp}{1 - sq} \bigg|_{s=1}$$

$$= \frac{2qp}{(1 - sq)^3} \bigg|_{s=1} = \frac{2q}{p^2} \tag{15.25}$$

We can now manipulate Equations (15.19) and (15.25) to obtain the exact expression for the variance:

$$E(X^2) = E(X(X-1)) + E(X) = \frac{2q}{p^2} + \frac{1}{p}$$

$$= \left(\frac{2qp + p^2}{p^2} \right) \tag{15.26}$$

The variance is

$$E(X^2) - (E(X))^2 = \frac{2q}{p^2} + \frac{1}{p} - \frac{1}{p^2}$$

$$= \frac{2q + p - 1}{p^2} = \frac{q + (q + p - 1)}{p^2} = \frac{q + (1 - 1)}{p^2}$$

$$= \frac{q}{p^2} \tag{15.27}$$

Note, in the above simplification $p + q = 1$.

15.2.6 Probability-Generating Function of Negative Binomial Random Variables

Consider an experiment consisting of N repeated trials in which each trial consists of two possible outcomes. The trials are independent. The outcomes are either a success with the same probability p or failure with the same probability q. The experiment is repeated until r successes are recorded. The value of r is specified a priori. The random variable X involved in this type of experiment is called a negative binomial random variable with a negative binomial distribution. The distribution is also sometimes called Pascal distribution.

An example of a negative binomial distribution is when a coin is tossed repeatedly. Suppose we specify a priori that 7 successes are desired, or that we need the coin to land on its head 7 times during the course of the experiment. Since the coin is unbiased, each toss is independent of the previous toss. That we had a head previous toss does not in any way affect what the outcome of the next toss would be (it could be head or tail). The probability of a head in each toss is 0.5 and the probability of a tail in each toss is also 0.5. We continue the experiment until 7 heads are obtained in total.

As usual we represent a success with probability p and failure with probability q (where $q = 1-p$), the number of trials with N and the number of successes with r. This means out of N trials $N - r$ failures result. In the analysis of the distribution and probability-generating function therefore, what we have is the number of combinations of N trials in which r successes result.

Consider a country where by law it is legislated that each IVF doctor is allowed to achieve r pregnancies in one year. Some doctors may try many times (N_j) before achieving the r pregnancies with probability of success (p_j) while others may try fewer times to achieve the same result. We represent the random process involved with the negative binomial probability distribution as $b(N_j; r, p_j)$.

This process can be used to rank and rate IVF doctors in terms of successes and failures and to advise the general public on where to go to achieve quick success. The general distribution is

$$b(N_j; r, \mathrm{p}_j) = {}_{N_j-1}C_{r-1}\mathrm{p}_j^r q_j^{N_j - r} \tag{15.28}$$

For the rest of this section, only one random process is involved (implantation by one IVF doctor on a single female). Let us now push forward with the analysis of negative binomial processes or random variables. Each negative binomial random variable consists of several geometric random variables. Consider the negative binomial random variable $X = X_1 + X_2 + \cdots + X_k$, where each component of the random variable X is a geometric random variable with geometric distribution. For example

$$X_i \overset{\Delta}{=} Geom(p) \tag{15.29}$$

Is the number of tosses of a biased coin between the $(i - 1)$th and ith head (success). The coin is tossed until k heads (success) result. Let all the X_i geometric components of the process be independent. Therefore, the probability-generating function of the random variable (negative binomial) is a product of the k (geometric) probability-generating functions such that

$$G_X(s) = \prod_{k=1}^{K} G_{X_k}(s) = \left(\frac{sp}{1 - sq}\right)^k, \quad |s| < \frac{1}{q} \tag{15.30}$$

15.2.6.1 Negative binomial probability law

In this section, we seek to derive the binomial probability distribution law. We will apply a few well-known techniques in calculus to simplify this expression further. Consider the well-known power series expansion of the sum $(1 + x)^\alpha$, $x \in (-1, 1)$, where α is a real number given by:

$$1 + \frac{\alpha x}{1!} + \frac{\alpha(\alpha - 1)}{2!} x^2 + \cdots + \frac{\alpha(\alpha - 1) \cdots (\alpha - r + 1)}{r!} x^r + \cdots$$

so that

$$\binom{\alpha}{r} = 1 + \frac{\alpha}{1!} + \frac{\alpha(\alpha-1)}{2!} + \cdots + \frac{\alpha(\alpha-1)\cdots(\alpha-r+1)}{r!}$$

$$(15.31)$$

Then, we have the binomial series

$$(1+x)^{\alpha} = \sum_{r \geq 0} \binom{\alpha}{r} \cdot x^r \qquad (15.32)$$

Except for a sign change, this expression is similar to the probability-generating function for a negative binomial distribution. Therefore:

$$G_X(s) = \prod_{k=1}^{K} G_{X_k}(s) = (sp)^k (1-sq)^{-k}$$

$$= (sp)^k \sum_{r=0}^{\infty} \binom{-k}{r} (-sq)^r \qquad (15.33)$$

The term inside the summation sign can be rewritten as

$$\binom{-k}{r} = \frac{-k(-k-1)\cdots(-k-r+1)}{r!}$$

$$= (-1)^r \binom{k+r-1}{k-1} \qquad (15.34)$$

We now have all the component terms to write the PGF for the negative binomial distribution clearly as

$$G_X(s) = \sum_{r=0}^{\infty} \binom{k+r-1}{k-1} p^k q^r s^{k+r} = \sum_{n=k}^{\infty} \binom{n-1}{k-1} p^k q^{n-k} s^n$$

$$(15.35)$$

$$P(X = n) = \begin{cases} 0, & n < k \\ \binom{n-1}{k-1} p^k q^{n-k}, & n = k, k+1, \ldots \end{cases} \qquad (15.36)$$

This expression is called the **negative binomial probability** law. It is the probability mass function for the negative binomial random variable when n is equal to or more than k.

15.3 Applications of Probability-Generating Functions in Data Analytics

In general, the probability-generating function associates with a sequence $c_0, c_1, c_2, c_3 \ldots$ a series based on the values taken by a random variable X.

$$G(x) = c_0 + c_1 x + c_2 x^2 + \cdots$$

Consider a random variable X whose values are integers defined by probabilities of the form $P(X = n) = p_n$. The associated series called the probability-generating function is

$$G(x) = p_0 + p_1 x + p_2 x^2 + p_3 x^3 \cdots$$

The exponent of x indicates the value that the random variable assumes. The associated probabilities are the probabilities of the random variable when it takes values in the exponent.

15.3.1 Discrete Event Applications

15.3.1.1 Coin tossing
In a coin tossing experiment, an unbiased coin may turn up "head" or "tail". We attach the values 1 to heads $(p = 1/2)$ and 0 $(q = 1/2)$ to tails. Each of the two outcomes have equal probabilities which are $p = q = 1/2$. The probability-generating function is

$$G(x) = p_0 + p_1 x = \frac{1}{2} + \frac{x}{2} \tag{15.37}$$

Exercises

(a) Show that the probability-generating function for tossing two unbiased coins is $G(x) = \left[\frac{1}{2} + \frac{x}{2} \right]^2$

(b) Suppose three coins are tossed, list the possible outcomes and show the probability-generating function is $2^{-3}(1 + x)^3$. (Hint: the outcomes are 000, 001, etc.)

(c) What is the probability-generating function for tossing N unbiased coins?

15.3.1.2 Rolling a die
A die has six faces. In an unbiased die, each face has equal probability of showing up which is 1/6. The probability-generating function for tossing

a die is:

$$G(x) = p_1 x + p_2 x^2 + p_3 x^3 + p_4 x^4 + p_5 x^5 + p_6 x^6$$

$$= \frac{x}{6} + \frac{x^2}{6} + \frac{x^3}{6} + \frac{x^4}{6} + \frac{x^5}{6} + \frac{x^6}{6} \qquad (15.38)$$

Exercise: (a) Two unbiased die are tossed, write an expression for the probability-generating function.

Solution:

$$G(x) = G_1(x)G_2(x) = [G_1(x)]^2$$

$$= \left[\frac{x}{6} + \frac{x^2}{6} + \frac{x^3}{6} + \frac{x^4}{6} + \frac{x^5}{6} + \frac{x^6}{6} \right]^2$$

$$= 6^{-2} [x^2 + 2x^3 + 3x^4 + 4x^5 + 5x^6 + 6x^7$$

$$+ 5x^8 + 4x^9 + 3x^{10} + 2x^{11} + x^{12}]$$

Notice that tossing the two die provide two independent events and therefore the probability-generating function is the product of equal probability-generating functions

$$G_1(x) = G_2(x) = G(x).$$

15.3.2 Modelling of Infectious Diseases

Recently, J. C. Miller [1] has published a primer on the use of probability-generating functions in infectious disease modelling. This section derives input for the excellent and timely work by Miller. We begin by examining the so-called "discrete-time early spread of a simple disease" [1] during the early time. This was defined in [1] with the probability-generating function. We use in this section lower case $G(\cdot)$ to refer to applicative-generating functions as discussed in this section. The discrete-time early spread probability-generating function is:

$$G(y) = \sum_{i=0}^{\infty} p_i y^i \qquad (15.39)$$

We use p_i as the probability of causing i infections, the so-called "offspring" before the person recovers.

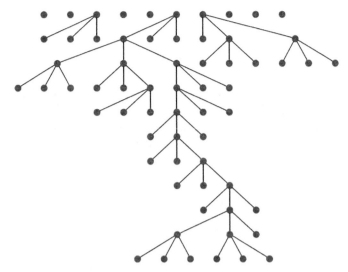

Figure 15.1 A sample of disease outbreaks. A sample of 10 outbreaks starting with a bimodal distribution having $R_0 = 0.9$ in which 3/10 of the population causes 3 infections and the rest cause none. The top row denotes the initial states, showing each of the 10 initial infections. An edge from one row to the next denotes an infection from the higher node to the lower node. Most outbreaks die out immediately [1].

The expected number of infections caused by an infector at the early stage in the outbreak of the disease is given by the expression

$$E(y) = R_0 = \left.\frac{d}{dy}G(y)\right|_{y=1} = \sum_i ip_i = G'(1) \qquad (15.40)$$

In Figure 15.1, three out of a population of 10 (0.3) infections cause 3 infections. One of the infections die out quickly. The second persisted for quite a while to ten more generations of infections before dying out. The third persisted for two generations of infections before dying out.

15.3.2.1 Early extinction probability

The objective of extinction theory and in fact extinction is to find the probability that a disease outbreak from one or more infected persons dies out. Assume that an infected person creates an independent set of offspring k chosen from a random variable with a distribution having probability-generating function $g(y)$. Let α be the probability of extinction if the epidemic starts from one person. Then

$$\alpha = \sum_k p_k \hat{\alpha}^k = G(\hat{\alpha}) \tag{15.41}$$

where $\hat{\alpha}$ is the probability that, in isolation, an offspring of an infected person will not cause an epidemic.

The probability that an outbreak caused by a single infected person goes extinct has probability α, which satisfies the PGF

$$\alpha = G(\alpha) \tag{15.42}$$

The coefficients of a PGF can be calculated if the actual value of the PGF is known at any point. This is given in [1]. Let

$$G(x) = \sum_n G_n x^n \tag{15.43}$$

The probabilities p_n are coefficients of a polynomial in x of power n. They may be obtained by taking the derivative of the PGF and evaluating it at $x = 0$ and dividing by n! (n factorial). Formally this is

$$p_n = \frac{1}{n!} \frac{d^n G(x)}{dx^n}\bigg|_{x=0} \tag{15.44}$$

This result holds provided the function $G(x)$ has a Taylor series expansion.

It should be understood that not every value of x satisfies this equation $x = G(x)$ [1]. More than one value of x will satisfy this equation, of which $1 = g(1)$ is a solution. There is another solution provided $\Re_0 = G'(1) > 1$.

In general, extinction probability in the limit is given by the expression

$$\alpha = \lim_{G \to \infty} \alpha_G; \quad \text{and} \quad \alpha_G = G(\alpha_{G-1}) \tag{15.45}$$

This results in a non-zero epidemic probability when we start from $\alpha_0 = 0$. The non-zero epidemic probability is

$$\Re_0 = G'(1) = \sum_k k \cdot p_k > 1 \tag{15.46}$$

15.3.2.1.1 *Models of extinction probability*

Two probability-generating function models are used in the study of probability of epidemiological extinction. The first is on infectious individual, creating the so-called offspring leading to the probability-generating function. This is seen from Figure 15.1 just after the initial infection. The second PGF "applies to the entire infectious class $I(t)$ at time t" [4]. The offspring PGF is given by the expression

$$G(x) = \sum_{j=0}^{\infty} p_j x^j, \quad x \in [0,1] \tag{15.47}$$

This is the probability of one individual j generating other j infections.

Properties
Several properties of this model can be itemized as

(1) $G(1) = 1$
(2) The mean number of offspring created by one infector is

$$G'(1) = \sum_{j=1}^{\infty} j p_j$$

(3) The probability-generating function for offspring infectors is given by the expression

$$G(x) = p_r + p_i x^2 = \frac{\gamma}{\beta+\gamma} + \frac{\beta}{\beta+\gamma} x^2; \quad x \in [0,1]$$

where γ is defined as a constant rate of recovery and β is the rate of transmission. In the equation p_r is the probability of recovery and p_i is the probability of an infected person infecting another person. $p_r + p_i = 1$. The power of x is the number of new infected people by the one infected person. The expectation from this expression is

$$G'(1) = \frac{d}{dx}G(x)\Big|_{x=1} = \frac{2\beta}{\beta+\gamma} \tag{15.48}$$

$G'(1) > 1$ *if* $R_0 > 1$. Further information on this model are found in [4]. We do not pursue the second model further as it leads naturally to the Markov model. Further reading on the topic is found in [1] and [4].

References

[1] Joel C. Miller, "A Primer on the Use of Probability Generating Functions in Infectious Disease Modeling", https://arxiv.org/abs/1803.05136, August 2018.

[2] Ping Yan, "Distribution theory, stochastic processes and infectious disease modelling", Mathematical Epidemiology, pp. 229–293, 2008.

[3] Herbert S. Wilf, Generating functionology, A K Peters, Ltd, 3rd edition, 2005.

[4] Linda J.S. Allen, "A primer on stochastic epidemic models: Formulation, numerical simulation, and Analysis", Infectious Disease Modelling, March, 2017, pp. 1–30.

16

Digital Identity Management System Using Artificial Neural Networks

Johnson I. Agbinya[1] and Rumana Islam[2]

[1]Melbourne Institute of Technology, Australia
[2]University of Technology, Sydney, Australia

16.1 Introduction

The need for cyber security is leading to a highly complex identity management environment. The complexity is due to the diversity of protocols, networks and devices as well as user requirements. The combination of distributed attributes and identities is often not associated across identity management systems and carries different amounts of security data. This causes difficulties due to multiple identities and raises the need for an efficient design of an identity management system. The multiplicities of approaches to and techniques for identity management, while dealing with the present challenges, have introduced new problems, especially the concern about privacy. Biometrics technologies nowadays are typically considered a vital component of the security requirements and are strongly attached to the basis of highly secure identification and authentication solutions. This chapter describes the fundamental concepts of identity authentication and identity management systems. We also propose and implement a multi-modal Digital Identity Management System (IDMS) integrating pseudo metrics, physical metrics and biometrics techniques based on artificial neural networks.

16.2 Digital Identity Metrics

With the rapid development of information and network technology, governments, colleges and enterprises are employing more and more information systems. As a result, they face a series of important technical challenges, one of them being the management of various user identities. Identity convergence however increases additional security and privacy issues. It is obvious that the need for identity that would provide complete privacy is vital.

Digital identity is a major objective when making services and resources available through computer networks. Understanding how digital network identification is different from paper-based identification systems is necessary and significant element in appropriate design of identity system [1]. Digital identity has diverse meanings based on different usage. Subenthiran has defined digital identity as "the representation of a human identity that gives an individual the power to control how they interact with other machines or people and share their personal information via Internet" [2].

Through the authentication process of identity, individuals can access to the different services. There are three types of metrics which combine a multitude of credentials that can be used for authentication [1, 3]:

- Pseudo metrics: "something you know", such as a PIN or password.
- Physical metrics: "something you have" likes a passport.
- Biometrics: "something you are" likes a voice or face recognition.

Successful solutions depend on the complex processes for utilization of these authentication parameters. Using pseudo or physical metrics have some deficiencies that restrict their applicability [4]. The main problem is that they may be forgotten or taken and used by unauthorized users. Biometric recognition systems however characteristically promise a greater convenience, comfort and security for their users. Physical human characteristics are hard to be counterfeited, stolen or misplaced; they are unique to every individual [1, 4, 5]. Although, as with any technology, biometric also has its drawbacks. An individual might feel that scanning and other methods of biometrics identification are intrusions to their privacy. For example, fingerprinting was associated with criminals for a long time and therefore having their fingerprint taken might make some people uncomfortable [1]. Furthermore, user's privacy can also be invaded if a cross-matching between different biometric databases is performed overtly, in order to track the enrolled subjects using their personal biometric characters [5]. Therefore, unauthorized and overt access to the stored biometric data is probably one of the most dangerous threats regarding user privacy and security.

The aim is to aid implementation of a multi-modal Digital Identity Management System (DIMS) in distributed mobile computing environment, which covers the accurate individual recognition and security via a combination of different types of authentication.

16.3 Identity Resolution

The critical challenge in authenticating an individual in an identity management system is to evaluate, compare and correlate all the accessible attributes of the user. A matching technique known as identity resolution is described in [6]. This method was originally designed for identity matching issues at Las Vegas casinos. It uses a deterministic technique based on expert rules in combination with a probabilistic component to determine generic values for identifiers.

The Artificial Intelligence Lab at University of Arizona developed algorithms that automatically recognize fake identities to aid police and intelligence investigations. A criminal record comparison method to address the problem of identifying fraudulent criminal identities is proposed in [7]. A supervised training process is employed to establish a proper disagreement threshold for matching, which is the total disagreement between two records is defined as the summation of the quantized disagreements between their constructing fields. Consequently, this method needs a set of training data, fake identities on online social network sites such as Facebook, Twitter and Instagram to mention a few. When combined with fake news, fake identities pose serious problems in crime detection, accurate polls in national elections and how much trust is to be attributed to presented identities in public places such ports of entries including airports.

To deal with the same problem, a probabilistic Naive Bayes model is proposed in [8]. This model utilizes the same features for matching identities as in [7], but a semi-supervised learning method is used instead, which reduces the effort for tagging training data. Phiri and Agbinya suggest a system for management of digital identities using techniques of information fusion in [9]. In our implementation, this technique is exploited.

16.3.1 Fingerprint and Face Verification Challenges

Biometrics has strong appeal in digital identity management systems because users no longer have the worries of remembering their passwords or losing their plastic cards. These types of authentications are usually permanent and are extremely difficult to forge.

In general, the process of using any type of biometric system for recognition purposes requires several steps, which include capturing of the raw biometric data, pre-processing and feature extraction, a recognition or matching algorithm and performance measures. This section will cover these processes, but it focuses only on how fingerprints and face verifications are undertaken.

16.3.1.1 Fingerprint

In Automatic Fingerprint Identification Systems (AFIS), fingerprint matching is the most important problem. There are two methods which have generally been used in AFIS for their matching algorithms: minutiae-based and texture-based matching [10].

The minutia-based algorithm is widely used for fingerprint authentication. It focuses on the endings of ridges and bifurcations. Consequently, the central area in fingerprint image is very important and this algorithm keenly relies on the quality of the input images [10, 11]. For example, in non-ideal conditions, a fingerprint image may have inadequately defined features caused by inking, scars, etc. In view of the fact that in this method image excellence has the greatest impact on matching performance, it is necessary to combine this algorithm with an image enhancement algorithm [12, 13].

Numerous studies have been carried out on parameters optimization. Some of the more recent researches are on using genetic algorithm [14], fuzzy systems [15], wavelet statistical features [16], geometric prediction [17], radial basis function neural network (RBFNN) [18], principal component analysis (PCA) [11] and convex hulls [19]. In this work, a minutia-based algorithm is used for fingerprint matching assuming that there is access to fairly clear fingerprints.

16.3.1.2 Face

As one of the most successful applications of image analysis and understanding, face recognition has recently received significant attention, especially in the past several years. At least two reasons account for this trend: the first is the wide range of commercial and law enforcement applications, and the second is the availability of feasible technologies after approximately 30 years research.

Even though current machine recognition systems have reached a certain level of maturity, their success is limited by the conditions imposed by many real applications. For example, recognition of face images acquired in an outdoor environment with changes in illumination, pose, scale, facial expression and background [20, 21].

The basic face recognition process consists of two computational stages. The first stage is feature extraction and classifying the facial images based on the chosen features is the second stage [22].

A huge number of techniques have been applied to facial feature extracting and can be divided into two types: segmentation and non-segmentation. The first one is geometric feature-based method, which is quite time consuming, mainly for large databases, but requires less data input [20, 22]. Principal component analysis (PCA) and wavelet coefficient are the typical methods of non-segmentation category. In our implementation, in order to extract the features, a geometric feature-based method has been employed for its accuracy.

16.4 Biometrics System Architecture

The biometric system architecture consists of four modules (Figure 16.1): enrollment phase, feature extraction, verification and identification. In this chapter, we mostly have considered verification and the identification is not in the scope of this work.

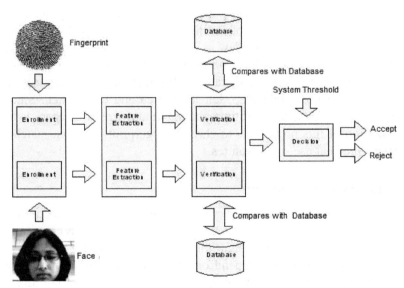

Figure 16.1   Biometric system structure.

Figure 16.2   Major fingerprint classes. (1) Whorl, (2) Right Loop, (3) Left Loop, (4) Arch, (5) Tented Arch.

16.4.1 Fingerprint Recognition

A number of fingerprint recognition software development kits (SDK) have been created by many companies to allow developers and integrators to obtain and verify fingerprint data from fingerprint scanners. For this implementation in Figure 16.1, we have used GrFinger Fingerprint SDK, which is based on minutiae-based algorithm and has a matching speed of up to 35,000 fingerprints per second for our Digital Identity Management System.

The uniqueness of the fingerprint is identified based on the characteristics and relationships of bifurcations and endings in ridge or valley. In order to compare two fingerprints, a set of invariant and discriminating features are extracted from fingerprint image. Most verification systems providing a high security are based on minutiae matching. In minutiae-based algorithms, template are created from these points and stored as relatively small template data and only a small part of the finger image is needed for verification. Fingerprint classification is view as a coarse-level matching of fingerprints. The first classification of fingerprints was done by Sir Edward Henry, when he developed the Henry classification scheme [17]. It classifies features into five classes; whorl, right loop, left loop, arch and tented arch. Since each match takes some amount of time, the maximum size of database is limited. A solution of this problem is to classify fingerprint, and first matching the fingerprints that use small features.

16.4.2 Face Recognition

As has been discussed before, there are two main types of techniques used in face recognition, segmentation and non-segmentation. In this chapter, the first one which is a feature-based algorithm has been selected for face recognition. Feature-based algorithms depend on extracting, derivation and analysis to gain the required knowledge about facial features that are not affected by variations in lighting conditions, pose, and other factors which most face recognition systems face. The feature-based techniques extract and normalize a vector of geometric description of biometric facial components such as the eyebrow, thickness, nose anchor points, chin shape, etc. [23]. The vector is then mapped with, or matched against, the stored model face vectors.

There are numerous face recognition software development kits (SDK) on the market that have been supplied by many different companies for a price. They all offer similar features, but the main feature they all provide is they allow developers and integrators to enroll and verify facial images. The face recognition SDK that has been chosen in this chapter is called FACEVacs by a company called Cognitec. It is compatible and integrate able with the GrFinger Fingerprint SDK for Digital Identity Management System. The main function performed by the SDK software to process images as shown in Figure 16.3 are:

- **Face and eye Finder:** to locate the face on the image, a pyramid like structure is created on the original image. The image pyramid is made up of copies of the original image at different scales, hence representing the image at sets of different resolutions. Over each image in the pyramid, a mask is moved from pixel to pixel, which in turn at each specific section under the mask, a function is passed that accesses the likeness of the image of the face in database. If the score passes a certain point of similarity, then the presence of a face at that particular position is automatically assumed. From that given information, the position and size of the face, position and resolution of the original image can be calculated. You can estimate the eye position be assessing the location and position of the face. Within this estimated positions and its zone, a search for the exact eye positions are started. The search for the eyes and face are similar except for the resolution of the images in the pyramid is higher than the resolution at which the face was previously found. And finally, the highest scores for the eyes are taken from the pyramid mapping function and the as final estimates of the eye positions. [24].
- **Image quality check:** The quality of the face image is checked by processes.

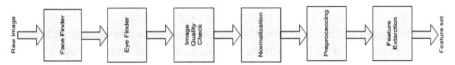

Figure 16.3 Feature set creation [24].

Figure 16.4 Eye locations found by the FaceVACS algorithm [24].

- **Normalization:** Centre of the eye at fixed positions within that image by extracting the image, which is turn is scaled and rotated in such a way that the result then becomes fix.
- **Processing:** The normalized image is preprocessed through a variety of standard techniques such as equalization, histogram, intensity normalization and a host of other techniques.
- **Feature extraction:** From the preprocessed image, distinctive features are then extracted from the image significant to distinguishing two or more individuals from each other.
- **Construction of the reference set:** A biometric template of facial features extracted from images in the enrollment phase.
- **Comparison:** This is used for verification purpose, the reference set of the individual in the image is processed for identification, the image is compared to all stored reference sets, and the individual with the largest comparison value is chosen, in all cases of verification is considered successful, the score then exceeds a certain threshold.

16.5 Information Fusion

The current management of digital identity authentication systems, which depends so much on a PIN number or a username and a password, has lead to an increase in online fraud since these credentials are easy to guess by hackers [25]. Multi-modal authentication, which involves combining a number of attributes in order to authenticate a user or a device, is considered as a possible solution to this problem. The process of combing these attributes is referred to as information or data fusion [26]. The application of information fusion technical systems requires mathematical and heuristic techniques from fields such as statistics, artificial intelligence, operations research, digital signal processing, pattern recognition, cognitive psychology, information theory and decision theory [27]. The aim of the information fusion engine is to compose the combined strength and correlation of the submitted attributes during multi-modal authentication. Information fusion is critical in the case of effective decision systems. However, it can be challenging because the technologies for utilizing the separate sources are not always appropriate or adequate for discovering value and intelligence from the fused data [28]. Artificial neural networks, fuzzy logic, Bayesian method, evolutional computation, hybrid intelligent systems and data mining technologies are considered as examples of the technologies that can be used to implement an information fusion technique [29]. In the next section, we discuss in detail about the artificial neural networks to implement the information fusion.

16.6 Artificial Neural Networks

The segment of artificial intelligence called artificial neural network (ANN) aims at emulating the function of the biological nervous system that makes up the brains found in nearly all higher life forms found on Earth [29]. Neural networks are made up of neurons and a neuron is made up of a core cell and several long connectors, which are called synapses [27]. These synapses show how the neurons are connected amongst themselves. Both biological and artificial neural networks work by transferring signals from neuron to neuron across the synapses. An artificial neural network is therefore an "information-processing paradigm that is inspired by the way biological nervous systems like the brain processes information" [29]. The key element of this concept is the novel structure of the information processing system. It is composed of a large number of highly interconnected processing elements (called neurons) working in unison to solve specific problems.

An artificial neuron commonly known as a perception is a device with many inputs and one output [29]. The neuron has two modes of operation, namely the training mode and the using mode [30]. In the training mode, the neuron can be trained to fire or not to fire, for particular input patterns. In the using mode, when a taught input pattern is detected at the input, its associated output becomes the current output. If the input pattern does not belong in the taught list of input patterns, the firing rule is used to determine whether to fire or not [30].

A trained artificial neural network is an expert in the category of information it has been given to analyse and can then be used to provide projections given new situations [29]. This makes it more useful when implementing information fusion. Once the network is trained and the network weights and threshold values are generated using a given range of training data to obtain a given set of targeted data, it can then work as an expert to compute the output of the information fusion based on the values given. The challenge when using artificial neural networks is how to compute the neuron weights and threshold values. Neural networks with their extraordinary ability to derive meaning from complicated or imprecise data can be used to extract patterns and detect trends that are too complex to be noticed by either humans or other computer techniques [30].

16.6.1 Artificial Neural Networks Implementation

Due to the complexness of artificial intelligent technologies, it is decided to use a multilayer artificial neural network to implement the information fusion engine. MATLAB has been used for neurons train.

The digital identification system implemented used six attributes as the input vector from the three groups: physical metrics, pseudo metrics and biometrics. It is important to have a reasonable number of attributes as too many attributes may slow down the system and discourage people from using it. While using too many attributes slows down the system and discourages users who have to submit a lot of credentials to access the services, too few attributes may compromise the required security of the system [31]. In this system, the email number, full name and race are used as the physical metrics. The password is used as the pseudo metrics and the face image and fingerprint as the biometric feature. The information fusion engine uses a set of six input vectors from three input sources.

- Input variables x1, x2, x3 form the vector for the physical metrics. This is then passed to the first hidden layer representing physical metrics and depicted by neuron 7.
- Input variable x4 forms the vector for pseudo metrics. This input is passed to the second hidden layer representing the pseudo metric grouping depicted by neuron 8.
- Input variable x5 and x6 form the biometrics vector and is fed into the third hidden layer representing the device metrics grouping and is depicted by neuron 9.
- Neuron 10 is the output neuron representing the output layer, which receives the inputs from the four hidden layers to calculate the network output.
- Y10 represents the overall combined weight of the submitted attributes. This is the value from the information fusion engine that is used in the multi-modal authentication system.

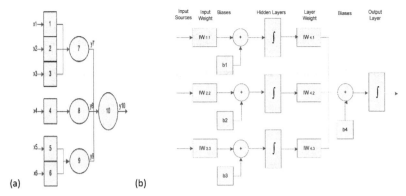

Figure 16.5 (a) Multilayer Artificial Neural Network [33]. (b) MATLAB design of Neural Networks [31].

A more detailed diagram representing the design in MATLAB is shown in Figure 16.5(b). The numbers on the left in the figure represent the number of input vectors. It shows the designed artificial neural network with three input sources, three input weights (IW), threshold values/biases (b), three hidden layers ($\int$), layer weights (LW) and one output layer [31].

16.7 Multimodal Digital Identity Management System Implementation

The DIMS will consist of several components both physical and conceptual, configured to optimise resources and maximise efficiency. Figure 16.6 shows the conceptual Digital Identity Modelling and Analysis architecture of the system with physical components located in the client tier of the model. The main components of the DIMS in terms of functional requirements are described in this section.

16.7.1 Terminal, Fingerprint Scanner and Camera

Terminal used to access the DIMS, this allows users to submit credentials to the system and allows the system to automatically detect device information from the terminal. One or more terminals are allowed to access the system, in conjunction with corresponding sets of scanners and cameras.

A finger scanning device is used to acquire fingerprints from the user, and in conjunction with a terminal and camera, submits the sets of attributes to the

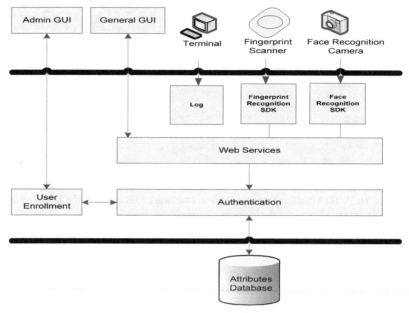

Figure 16.6 Conceptual Architecture of DIMS.

DIMS for authentication. Communication between the scanner and system is via a terminal's USB port.

A camera device is used for face recognition of the user, and in conjunction with a terminal and fingerprint scanner, submits the sets of attributes to the DIMS for authentication. Communication between the camera and system is via a terminal's USB port.

16.7.2 Fingerprint and Face Recognition SDKs

GrFinger Fingerprint SDK Biometric Recognition Library by Griaule is a component package responsible for the retrieval, verification, enrolment and identification of fingerprints. This component consists of libraries of classes to be used by the DIMS.

FaceVacs Face Recognition SDK by Cognitec is responsible for the retrieval and verification of face. This component consists of libraries of classes to be used by the DIMS.

16.7.3 Database

A secure database was developed in MySQL, which takes four user inputs and stores in the database. This current system is the first stage of the DIMS in which a secured database will be developed.

16.7.4 Verification: Connect to Host and Select Verification

The user connects to host and selects verification in the Digital Identify Management System. The user clicks on connect and is transferred to a page that verifies the user to give access to a service.

16.7.4.1 Verifying user

At enrolment the system user is required to submit the full details of the subjects which are full name, email address, race and password. The system user then scans the subject's finger and captures the image by clicking on Verify. The user wanting to gain access to the service is verified against the details of the user already stored during the enrolment stage.

16.7.4.2 Successful verification

General system user clicks on verify and all user inputs are matched against the details in the database. If invalid input is entered, a popup message

specifying which attribute(s) is invalid/or missing is presented. Depending on the security threshold set by the admin during the enrolment stage, a subject will be accepted/rejected as a verified person.

16.8 Conclusion

In this chapter, a multimodal DIMS based on fusion of two biometric traits, namely fingerprint and face recognition, as well as pseudo and physical metrics has been proposed and implemented. Six attributes were chosen and a multilayer neural network was then used to combine these attributes to obtain accurate recognition of the user during multi-modal authentication.

For additional security, more attributes can be added to the current system; however, that raises social/privacy factors since more information will require to be collected from the users. The aim of the DIMS is to provide accurate and efficient identification of individuals under a non-intrusive and uncontrolled environments.

References

[1] Chenggang, Z. and S. Yingmei 2009, 'Research about human face recognition technology', Test and Measurement, 2009. ICTM 09. International Conference.

[2] Subenthiram Sittampalam, "Digital Identity Modelling and Management", MEng. Thesis, 2005, UTS, Australia.

[3] Bala, D. 2008. 'Biometrics and information security', Proceedings of the 5th annual conference on Information security curriculum development, Kennesaw, Georgia.

[4] Wayman, J. L. 2008, 'Biometrics in Identity Management Systems', Security and Privacy, IEEE conference.

[5] Bhattacharyya, D., R. Ranjan, et al. 2009, 'Biometric authentication techniques and its future possibilities'.

[6] J. Jonas, "Threat and fraud intelligence, las vegas style," Security and Privacy Magazine, IEEE, vol., pp. 28–34, 2006.

[7] G. A. Wang, H. Atabakhsh, T. Petersen, and H. Chen, "Discovering identity problems: A case study." in LNCS: Intelligent and Security Informatics. Spring Berlin/Heidelberg, 2005.

[8] G. A. Wang, H. Chen, and H. Atabakhsh, "A probabilistic model for approximate identity matching," in Proceedings of the 7th Annual

International Conference on Digital Government Research, DG.O 2006, San Diego, California, USA, May 21–24, 2006, J. A. B. Fortes and A. Macintosh, Eds. Digital Government Research Center, 2006, pp. 426–463.

[9] J. Phiri and J. Agbinya, "Modelling and Information Fusion in Digital Identity Management Systems," in Networking, International Conference on Systems and International Conference on Mobile Communications and Learning Technologies, 2006. ICN/ICONS/MCL 2006. International Conference on, 2006, pp. 181–187.

[10] Bey, K. B., Z. Guessoum, et al. 2008, 'Agent based approach for distribution of fingerprint matching in a metacomputing environment' Proceedings of the 8th international conference on New technologies in distributed systems, Lyon, France.

[11] Wang, Y., X. Ao, et al. 2006, 'A fingerprint recognition algorithm based on principal component analysis', TENCON 2006. 2006 IEEE Region 10 Conference.

[12] Changlong, J., K. Hakil, et al. 2009, 'Comparative assessment of fingerprint sample quality measures based on minutiae-based matching performance, Electronic Commerce and Security, 2009. ISECS '09, Second International Symposium.

[13] Zhi, Y., W. Jiong, et al. 2009, 'Fingerprint image enhancement by super resolution with early stopping', Intelligent Computing and Intelligent Systems, 2009. ICIS 2009. IEEE International Conference.

[14] Scheidat, T., A. Engel, et al. 2006, 'Parameter optimization for biometric fingerprint recognition using genetic algorithms', Proceedings of the 8th workshop on Multimedia and security, Geneva, Switzerland.

[15] Lopez, M. and P. Melin 2008, 'Topology optimization of fuzzy systems for response integration in ensemble neural networks: The case of fingerprint recognition', Fuzzy Information Processing Society, 2008. NAFIPS 2008, Annual Meeting of the North American.

[16] Nikam, S. B. and S. Agarwal 2008, 'Level 2 features and wavelet analysis based hybrid fingerprint matcher', Proceedings of the 1st Bangalore Annual Compute Conference, Bangalore, India, ACM.

[17] Ghosh, S. and P. Bhowmick 2009, 'Extraction of smooth and thin ridgelines from fingerprint images using geometric prediction', Advances in Pattern Recognition, 2009. ICAPR '09, Seventh International Conference.

[18] Jing, L., L. Shuzhong, et al. 2008, 'An Improved fingerprint recognition algorithm using EBFNN', Genetic and Evolutionary Computing, 2008. WGEC '08, Second International Conference.

[19] Chengming, W. and G. Tiande 2009, 'An efficient algorithm for fingerprint matching based on convex hulls', Computational Intelligence and Natural Computing, 2009. CINC '09. International Conference.

[20] Singh, S. K., V. Mayank, et al. 2003, 'A comparative study of various face recognition algorithms (feature based, eigen based, line based, neural network approaches)', Computer Architectures for Machine Perception, 2003 IEEE International Workshop.

[21] O'Toole, A. J., H. Abdi, et al. 2007, 'Fusing face-verification algorithms and humans' systems, Man, and Cybernetics', Part B: Cybernetics, IEEE Transactions on 37(5): 1149–1155.

[22] Jie, L., L. Jian-Ping, et al. (2007). Robust face recognition by wavelet features and model adaptation. Wavelet Analysis and Pattern Recognition, 2007. ICWAPR '07. International Conference.

[23] S. S. Mohamed, Aamer, et al. 2006. Face Detection based on Skin Color in Image by Neural Networks. [Online] 2006. [Cited: November 15, 2007.] http://www.istlive. org/intranet/school-of-informatics-university-of-bradford001-2/face-detection-basedon-skin-color-in-image-by-neural-networks.pdf.

[24] Cognitec. 2007. [Online] 2007. http://www.cognitec-systems.de/.Consor tium, The Unicode. 2003.

[25] Sabena, F., A. Dehghantanha, et al, 'A review of vulnerabilities in identity management using biometrics', Future Networks, 2010. ICFN 10. Second International Conference.

[26] Hall D., "Mathematical Techniques in Multisensor Data Fusion", Artech House, Boston, MA, 1992.

[27] Silk Roadpublications,"IDSDataFusion",SilkRoad Inc. http://www.silk road.com/papers/html/ids/node3.html, 2005.

[28] CSI. 2007. Advanced Technologies for Multi-Source Information Fusion. [Online] 2007. [Cited: May 10, 2007.] http://www.essexcorp.com/fusion.pdf.

[29] Micheal Negnevitsky, "Artificial Intelligence", Addison Wesley, www.booksites.net, ISBN 0-201-71159-1, 2002.

[30] Christos Stergiou and DimitriosSiganos, "NeuralNetworks", http://www.doc.ic.ac.uk/~nd/surprise_96/journal/vol4/cs11/report.html, 2005.

[31] Phiri, Jackson. 2007. Digital Identity Management System. South Africa: University of Western Cape, 2007. Unicode Standard, Version 4. [Online] 2003.

17

Probabilistic Neural Network Classifiers for IoT Data Classification

Tony Jan

Melbourne Institute of Technology, Sydney, Australia
E-mail: tjan@mit.edu.au

17.1 Introduction

The Internet of Things (IoT) is promised to create a new era of communication and business opportunities with the highly interconnected small devices. Currently, there are 9 billion interconnected devices and it is expected to reach 24 billion devices by 2020 [1]. According to the authority, this amounts to $1.3 trillion revenue opportunities for mobile network operators alone spanning vertical segments such as health, automotive, utilities and consumer electronics [1].

Since IoT will be among the greatest sources of new data, data science will make a great contribution to make IoT applications more intelligent. Data science is the combination of different fields of sciences that uses data mining, machine learning and other techniques to find patterns and new insights from data [2].

IoT devices generate enormous quantities of data that need to be aggregated and analyzed in real time [3]. The heterogeneous, high-velocity, and dynamic nature of IoT data poses significant challenge to the machine learning and pattern recognition community [4].

Identifying patterns from sample data is a timeless and useful exercise. Most of the pattern recognition algorithms are based on estimation of probabilistic density function (PDF) which models the output space with an estimate of probability of output class membership [5]. The estimation of PDF may be a challenge if a parametric model does not represent the underlying pattern well (which is likely the case in IoT data modelling). For such cases, a data-centric non-parametric kernel-based approximation such as probabilistic neural network (PNN) may be useful [6].

The PNN by Specht [6, 7] is based on well-established statistical principles rather than heuristic approaches. PNN is closely related to Bayesian decision strategy and nonparametric kernel-based estimator of PDF [7].

The mathematical finesse and easier visualization of PNN is flaring a renewed interest in its application to modern day data mining applications [12, 13]. The variants of PNN are well tested in heterogeneous and dynamic data modelling and they are considered suitable for IoT data modelling [22, 23].

In this chapter, we review the fundamental mechanisms of PNN and its variants for IoT data modelling.

17.2 Probabilistic Neural Network (PNN)

The PNN is based on well-established statistical principles such as Bayesian decision strategy and nonparametric kernel-based estimator of PDF [7].

PNN has both advantages and disadvantages. The advantages of PNN include a fast training (faster than back-propagation by approximately five orders of magnitude); guarantee to approach the Bayes optimal boundaries and a single smoothing factor [8]. Moreover, the output generated by PNN is in the form of probabilistic representation of class membership offers a meaningful interpretation of the outputs. The disadvantage of PNN is its demanding memory consumption because it actually includes as many RBFs as the number of training points.

This computational complexity may no longer an issue because of recent development in the variants of PNN utilizing vector quantization of its input space. Intelligent segmentation followed by employment of Adaboosting has shown the useful performance of PNN is modelling complex and dynamic problem with reasonable computational complexity [11, 13]. In addition, we are aided by more powerful computing systems.

The major idea of PNN is to group the points within an area of high probability density into clusters. More specifically, simple functions such as spherical Gaussians are associated with each training point and summed together to estimate the overall PDF. In theory, a sufficient number of training points can produce a good estimate for a true PDF.

Assume that there are p training vectors x_j and N classes C_i in the output space. The task for a PNN in this problem is to compute the PDF $g_i(x)$ for each output class. The highest $g_i(x)$ value is chosen to determine the class decision for an unknown vector x. In particular, the estimated PDF for a class C_i is given in Equation (17.1):

$$g_i(x) = \sum_{j=1}^{M} exp\left(\frac{Z_{ij} - 1}{\delta^2}\right) \qquad (17.1)$$

where

d: Single smoothing parameter chosen during network training.
M_i: Number of representative sample vectors in each class C_i.
Z_{ij}: The normalization of an input vector x_j with respect to class C_i.

From Figure 17.1, the input units are the distribution points for the input vectors. For each class C_i in the output space, we have M_i pattern units and a summation unit. Each pattern unit represents a training vector x_{ij} (input vector x_j with respect to class C_i) and performs the operation $exp(\frac{Z_{ij}-1}{\delta^2})$. These values from patterns units are summed to produce $g_i(x)$ for each class.

The only parameter that needs to be adjusted during the training phase is the single smoothing factor δ, which is the common radial deviation of all the Gaussian functions [6]. This parameter needs to be selected such that a reasonable degree of overlapping between adjacent RBFs is obtained. There are 2 possible ways to find the appropriate δ.

The first approach assumes that the class membership probabilities in the training data are the same as the actual probabilities in the population called

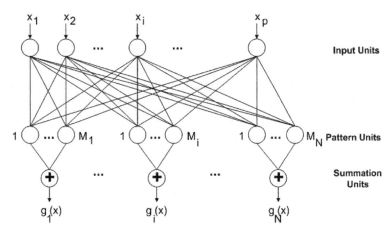

Figure 17.1 Architecture of probabilistic neural network [6].

prior probabilities [6]. If the prior probabilities are given, the network can train faster by adjusting its weighs to produce the class probabilities close to these values. In practice, noisy data can produce misclassifications and some of these errors are more serious than the others.

The second method is based on the loss factors reflecting the cost of misclassifications [6]. A fourth layer which contains a cost matrix is added after the output layer into the PNN. The network parameters which result in the lowest cost will be selected.

17.3 Generalized Regression Neural Network (GRNN)

Specht proposed GRNN [7] which operates in a similar manner to PNN (based on Bayesian estimation techniques) but is used for classification rather than regression. For example, given a set of input vectors with a known distribution, the task of classification is to produce the distribution function for the corresponding output space. According to the Bayesian theorem, the conditional distribution of the output space y for given input space x is given in Equation (17.2):

$$p(y|x) = \beta p(\gamma) p(x|y) \qquad (17.2)$$

where

$p(y)$: Prior distribution for y
$p(x|y)$: Conditional probability distribution of x given y
β: A coefficient that depends only in x but not y.

If the density $p(x|y)$ is not known, it must be estimated a *posteriori* from a set of sample observation x and y. This density can be alternatively computed non-parametrically by using a Gaussian Parzen estimator or kernel [7, 8].

Similar to PNN, GRNN uses a single common radial bias function kernel bandwidth δ that is tuned during the learning phase. In particular, an optimal δ will produce the lowest learning mean squared error (MSE). The following is the general form of GRNN, which is similar to the equation proposed by Nadaraya [14] and Watson [15]:

$$\hat{y}(\underline{x}) = \frac{\sum_{n=1}^{NV} y_n f_n(\underline{x} - \underline{x}_n, \delta)}{\sum_{n=1}^{NV} f_n(\underline{x} - \underline{x}_n, \delta)} \qquad (17.3)$$

With Gaussian function $f_n(\underline{x}) = \exp\left(\frac{-(\underline{x} - \underline{x}_i)^T (\underline{x} - \underline{x}_i)}{2\delta^2}\right)$

where

$\underline{x}$: Input vector (underline refers to vector)
$\underline{x}_n$: All other training vectors in the input space
δ: Single smoothing parameter chosen during network training
y_n: Scalar output related to $\underline{x}_n$
NV: Total number of training vectors

The Gaussian function can be rewritten in a more compact form:

$$f_i(d_i, \delta) = exp\left(\frac{-d^2}{2\delta^2}\right) \tag{17.4}$$

where $d_i = ||x - \underline{x}_i|| = \sqrt{(x - \underline{x}_i)^T (x - \underline{x}_i)}$ is the Euclidian distance between vector x and $\underline{x}_n$.

From the above GRNN architecture, each input vector has an associated equal-size Gaussian function and a corresponding scalar output. The Gaussian function is applied to the Euclidian distances of an input vector to all other vectors in the input space. This task is accomplished by the input

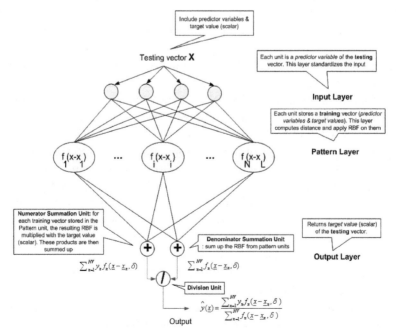

Figure 17.2 Architecture of generalized regression neural network [7].

and patterns units. This consideration of all vectors in the whole input space, in fact, causes a high computational cost for the system.

The Equation (17.3) is actually *a statistical non-parametric regression* method based on lots of data and Bayesian PDF estimation. As the number of data points goes to infinity, the method becomes independent of the basis function and is therefore non-parametric. However, if small data sets are used, the type and characteristics of the basis function has an effect on the regression. Hence, GRNN can be considered as semi-parametric in this case [7]. We discuss the semi-parametric nature of GRNN in later sections.

In many applications, GRNN demonstrates fairly high accuracy. However, it is computationally expensive as well as sensitive to the selection of variances for its Gaussian functions [8].

17.4 Vector Quantized GRNN (VQ-GRNN)

For some real-life complicated classification problems, artificial neural networks (ANNs) perform well in comparison with other approaches. However, ANNs often require large computational power. Therefore, in order to simplify the current ANNs without losing their non-linearity, some methods can be implemented to approximate non-linear models with acceptable predictive accuracy while reducing computational complexity to a reasonable level. The vector-quantized general regression neural network (VQ-GRNN), also known as the modified probabilistic neural network (MPNN), was introduced by Zaknich [8] for application to general signal processing and pattern recognition problems. VQ-GRNN is a generalization of Specht's probabilistic neural network (PNN) [6] and is related to Specht's general regression neural network (GRNN) [7] classifier. In particular, this method generalizes GRNN by assigning the clusters of input vectors rather than each individual training case to radial units. It has been proven to be able to improve the learning speed and efficiency by reducing the network complexity [8, 9].

When the number of data points in the training set is much larger than the number of degrees of freedom of the underlying process, we are constrained to have as many radial basis functions as the data points presented. Therefore, the classification problem is said to be overdetermined. Consequently, the network may end up fitting misleading variations due to noise in the input data, thereby resulting a degraded generalization performance [8]. GRNN normally suffers from this problem. In fact, GRNN is very computationally expensive because it incorporates each and every training example $(\underline{x}_i \rightarrow y_i)$

into its architecture. In order to overcome this problem, VQ-GRNN generalizes GRNN by quantizing the data space into clusters and assigning a specific weight to each of these clusters. Figure 17.3 is the architecture of VQ-GRNN [8].

In this structure, the Euclidian distances from the input vector $\underline{x}$ to the clusters within the input space $(c_1, \ldots, c_M)$ are computed. The Gaussian function is then applied to these distances. After that, two summing units are computed. The first one is the summation of the Gaussian functions while the second one is derived by adding the products of cluster size Z_i, the associated

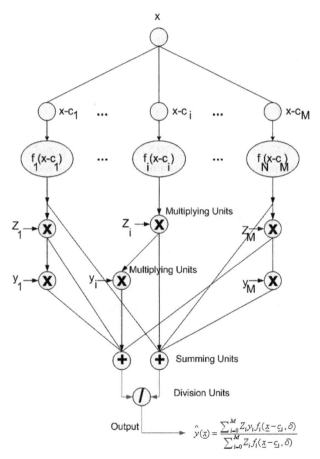

Figure 17.3 Architecture of vector quantized GRNN (VQ-GRNN) [8].

scalar output and the Gaussian functions. Finally, these terms are fed into the division unit. The following is the summary of VQ-GRNN model.

If there exists a corresponding scalar output y_n for each local region (cluster) which is represented by a centre vector $\underline{c}_i$, then a GRNN can be approximated by a VQ-GRNN formulated [8, 9]:

$$\hat{y}(\underline{x}) = \frac{\sum_{i=0}^{M} Z_i y_i f_i(\underline{x} - \underline{c}_i, \delta)}{\sum_{i=0}^{M} Z_i f_i(\underline{x} - \underline{c}_i, \delta)} \tag{17.5}$$

With Gaussian function $f_i(\underline{x}) = exp\frac{-(\underline{x}-\underline{c}_i)^T(\underline{x}-\underline{c}_i)}{2\delta^2}$
where

$\underline{c}_i$ = centre vector for cluster i in the input space
y_i = scalar output related to $\underline{c}_i$
Z_i = number of input vectors x_j within cluster $\underline{c}_i$
δ = single smoothing parameter chosen during network training
M = number of unique centres $\underline{c}_i$

Equation (17.5) can be seen as the general formulation for both GRNN and VQ-GRNN. In other words, GRNN can be computed from this equation by assuming that each cluster contains only one input vector ($Z_i = 1$), the y_i are real values (the output space is not quantized), the centre vectors $\underline{c}_i$ are replaced by individual training vectors x_i and the number of clusters is equal to the number of individual input vectors ($M = NV$) [8, 9]. Comparing Equations (17.3) and (17.5), the only difference is that VQ-GRNN applies its computation on a smaller number of clusters of input vectors represented by centres vectors $\underline{c}_i$ rather than dealing with individual input vectors $\underline{x}_n$ [8]. This clustering relies on Gaussian characteristics in which the sum of multiple Gaussian functions within a cluster is approximated by a single Gaussian with magnitude of Z_j, provided that the individual functions are well concentrated (clustered) near the centres in the data space:

$$\sum_{i=0}^{Z_j} f_i(\underline{x} - \underline{x}_i, \delta) \approx Z_j f_j(\underline{x} - \underline{c}_j, \delta) \tag{17.6}$$

This estimation is illustrated in Figure 17.4.

There are two methods of vector quantization used to train VQ-GRNN. The first method, Method A, relies on grouping similar y_i in the output space and then associating them with the mean c_i of a local group of input vectors. In particular, it involves uniformly quantizing the noiseless

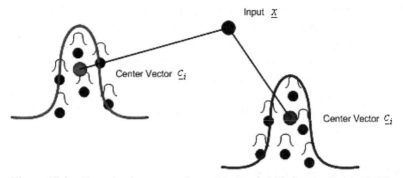

Figure 17.4 Clustering input space in modified probabilistic neural network [8].

desired y_i, separately grouping the y_i having positive and negative slopes in the waveform, and associating them with the mean of the input vectors mapping to each group [8]. This method produces a smaller network size, but it can only be used where one-to-one correspondence between small local regions in the input space and the output space is guaranteed (if this assumption is not guaranteed, the mean c_i of the input vectors x_j will not adequately represent a local region of those vectors). To overcome this problem, Method B is developed which involves uniquely clustering only those vectors in local hypercube regions of the input vector space that map to given quantized outputs by simply choosing suitable quantization parameters [8]. These parameters can be made coarser to further reduce the network size, but they should not be too coarse so that the quantization error becomes significantly greater than the expected residual noise in the network output. This latter method reduced to an efficient realization of a quantized version of GRNN.

Besides the Gaussian function, there are many other radial basis functions that can be chosen such as top-hat and triangular radial basis functions defined by Equations (17.7) and (17.8) respectively.

$$f_i(d_i, \delta) = \begin{cases} 1 & \text{if } d_i \leq \delta \\ 0 & \text{otherwise} \end{cases} \tag{17.7}$$

$$f_i(d_i, \delta) = \begin{cases} 1 - \dfrac{d_i}{\delta} & \text{if } d_i \leq \delta \\ 0 & \text{otherwise} \end{cases} \tag{17.8}$$

In practice, the Gaussian function is often found to be the function of choice for many problems unless there are computational constraints to be considered. The basis function is commonly chosen depending on minimal computational considerations in hardware realizations [8].

By using the vector quantization technique, the resulting VQ-GRNN model is always a semi-parametric version of the GRNN, which tends to smooth noisy data a little more than the GRNN. The VQ-GRNN retains the benefits of the GRNN with respect to generalization capabilities and ease of training by adjusting a single parameter δ, but the VQ-GRNN is always smaller in network size.

17.5 Experimental Works

There has been a large amount of research on IoT data classification from a limited set of IoT training data.

A classification of IoT data based on semantics is proposed in [16] and classification of IoT streaming data in [17]. A comprehensive review of machine learning for IoT data classification is provided in [18]. Other advanced machine learning models are proposed for IoT data mining but with mediocre results due to the real-time requirements of the IoT environment [19, 20].

We test our VQ-GRNN which is known to be of a semi parametric nature with adaptability for dynamic modelling (useful for IoT environment) on available IoT data.

First, data from the IoT devices are collected and filtered. The data are cleaned, and their initial classification is made to ensure the data are indeed from the IoT devices. The filtered data is then classified against the known patterns of the IoT usages. The unclassified data is confined for further analysis on its malicious signatures. If found not malicious, the data are further analyzed to define a new IoT service pattern.

In this experiment, we are targeting the classification of location-based services from handheld devices using the industry benchmark data offered by Mohammadi et al. [21].

The dataset was created using the RSSI readings of an array of 13 iBeacons in the first floor of Waldo Library, Western Michigan University. Data was collected using iPhone 6S. The dataset contains two sub-datasets: a labelled dataset (1420 instances) and an unlabeled dataset (5191 instances) [21].

The description of testing environment is extracted below:

> *The recording was performed during the operational hours of the library. For the labeled dataset, the input data contains the location*

(label column), a timestamp, followed by RSSI readings of 13 iBeacons. RSSI measurements are negative values. Bigger RSSI values indicate closer proximity to a given iBeacon (e.g., RSSI of −65 represent a closer distance to a given iBeacon compared to RSSI of −85). For out-of-range iBeacons, the RSSI is indicated by −200. The locations related to RSSI readings are combined in one column consisting a letter for the column and a number for the row of the position. Figure 2 depicts the layout of the iBeacons as well as the arrange of locations [21].

The data is fed to MATLAB Simulink with selected machine learning tools simulated. The experiment is a simple process of comparing the classification performance. There was no weighting to any service. The results show that the proposed model achieves good classification for both IoT service classification and security detection.

The experimental results demonstrate the usefulness of the proposed technique in IoT usage classification in comparison to the other machine learning models [19, 20].

17.6 Conclusion and Future Works

The paper proposed a compact and innovative machine learning tool for real-time IoT usage classification. The results demonstrate good use of PNN in dynamic modelling suitable for IoT environment. This work is

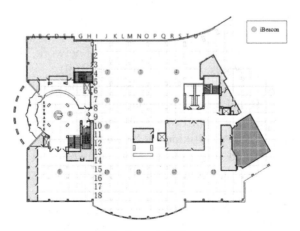

Figure 17.5   Layout of iBeacons [21].

useful paving the opportunities for the large-scale IoT data classification experiments.

The dynamic nature of IoT environments requires a case-by-case optimization of particular service type. The future work includes the use of an ensemble of simple classifiers that can be adapted in a semi parametric manner to accommodate the changing requirements of the IoT network.

References

[1] S. Vashi, J. Ram, J. Modi, S. Verma and C. Prakash, "Internet of Things (IoT): A vision, architectural elements, and security issues," 2017 International Conference on I-SMAC (IoT in Social, Mobile, Analytics and Cloud) (I-SMAC), Palladam, pp. 492–549, 2017.

[2] Mohammad Saeid Mahdavinejad, Mohammadreza Rezvan, Mohammadamin Barekatain, Peyman Adibi, Payam Barnaghi, Amit P. Sheth, Machine learning for Internet of Things data analysis: A survey, Digital Communications and Networks, 2017.

[3] In Lee, Kyoochun Lee, The Internet of Things (IoT): Applications, investments, and challenges for enterprises, Business Horizons, Volume 58, Issue 4, 2015, pp. 431–440.

[4] J. Manyika, M. Chui, B. Brown, J. Bughin, R. Dobbs, C. Roxburgh and A. H. Byers, Big Data: the Next Frontier for Innovation, Competition, and Productivity.

[5] E. Parzen, "On estimation of a probability density function and mode," Ann. Math. Stat., vol. 33, pp. 1065–1076, 1962.

[6] D. F. Specht, "Probabilistic neural networks," Neural Networks, vol. 3, pp. 109–118, 1990.

[7] D. F. Specht, "A general regression neural network," IEEE Transactions on Neural Networks, vol. 2, pp. 568–576, 1991.

[8] A. Zaknich, "Introduction to the modified probabilistic neural network for general signal processing applications," IEEE Transactions on Signal Processing, vol. 46, pp. 1980–1990, 1998.

[9] A. Zaknich and Y. Attikiouzel, "An unsupervised clustering algorithm for the modified probabilistic neural network," in IEEE International Workshop on Intelligent Signal Processing and Communications Systems, Melbourne, Australia, pp. 319–322, 1988.

[10] D. F. Specht, "Enhancements to the probabilistic neural networks," in Proceedings of the IEEE International Joint Conference on Neural Networks, Baltimore, MD, pp. 761–768, 1992.

[11] A. Zaknich, "An adaptive sub-space filter model," in International Joint Conference on Neural Networks (IJCNN), Portland, Oregon, USA, 2003, pp. 1464–1468.

[12] P. A. Kowalski and P. Kulczycki, 2017. Interval probabilistic neural network. Neural Computing and Applications, 28(4), pp. 817–834.

[13] T. Jan, Ada-boosted locally enhanced probabilistic neural network for IoT Intrusion Detection, 12th International Conference on Complex, Intelligent, and Software Intensive Systems (CISIS), Matsue, Japan, July 4–6th, 2018.

[14] E. A. Nadaraya, "On estimating regression," Theory Probability Applications, vol. 9, pp. 141–142, 1964.

[15] G. S. Watson, "Smooth regression analysis," Sankhya Series, vol. 26, pp. 359–372, 1964.

[16] M. Antunes, D. Gomes and R. L. Aguiar, Towards IoT data classification through semantic features. Future Generation Computer Systems, 2017.

[17] G. De Francisci Morales, A. Bifet, L. Khan, J. Gama and W. Fan, August. Iot big data stream mining. In Proceedings of the 22nd ACM SIGKDD International Conference on Knowledge Discovery and Data Mining (pp. 2119–2120), ACM, 2016.

[18] S. Bhatia and S. Patel, Analysis on different data mining techniques and algorithms used in IOT. Int. J. Eng. Res Appl, 2(12), pp. 611–615, 2015.

[19] Friedemann Mattern and Christian Floerkemeier, "From the Internet of Computers to the Internet of Things," in Lecture Notes In Computer Science (LNCS), Volume 6462, pp. 242–259, 2010.

[20] F. Chen, P. Deng, J. Wan, D. Zhang, A. V. Vasilakos and X. Rong, Data mining for the internet of things: literature review and challenges. International Journal of Distributed Sensor Networks, 11(8), p. 431047, 2015.

[21] M. Mohammadi and A. Al-Fuqaha, "Enabling Cognitive Smart Cities Using Big Data and Machine Learning: Approaches and Challenges," IEEE Communications Magazine, vol. 56, no. 2, 2018.

[22] T. Jan, "Neural Network Based Threat Assessment for Automated Visual Surveillance," in Proc. of International Joint Conference on Neural Networks IJCNN Budapest, Hungary, 2004.

[23] T. P. Tran and T. Jan, Boosted Modified Probabilistic Neural Network for Network Intrusion Detection, Proc. of IEEE International Joint Conference in Neural Networks (IEEE-IJCNN), Vancouver, BC, Canada, 2006.

18

MML Learning and Inference of Hierarchical Probabilistic Finite State Machines

Vidya Saikrishna[1], D. L. Dowe[2] and Sid Ray[2]

[1]Melbourne Institute of Technology, Melbourne Campus, Australia
[2]Monash University, Clayton, VIC, Australia

18.1 Introduction

Probabilistic Finite State Machines (PFSMs) are models that contain regularities and patterns of text data. The various sources generating such text data include a natural language corpus, a DNA sequence and an email text corpus. The models (PFSMs) are used for analysis of the text data such as classification and prediction.

This research work is focused on learning PFSM models from two classes of text data under a supervised learning environment. The model is a hypothesis and the information-theoretic Minimum Message Length (MML) principle is used to judge the goodness of the hypothesis in relation to prediction or classification, in different situations. In short, MML has been used as a technique to select among the competing PFSM models. We propose a novel approach for classification and prediction and apply it on two different problems.

The approach proposed for classification of text data under a two-class learning environment is the idea of learning hierarchical PFSMs or HPFSMs. This is an important contribution arising out of this research on accounts of its novelty and experimentally proven good results. We discuss a method of encoding HPFSMs and compare the code length of the HPFSM model with the traditional PFSM model. For a text data, if the inherent hierarchies are somehow found or assumed, then learning the HPFSM model for that text data is cheaper than learning the PFSM model for the same data. We show this comparison on at least two artificially generated HPFSM models and also on some publicly available datasets.

The conventional method of learning non-hierarchical PFSMs or simply PFSMs is extended to the case of learning Hierarchical PFSMs or HPFSMs in this chapter. The HPFSMs represent the behavior of PFSMs more concisely by identifying the inherent hierarchy in the data that is a sequence of tokens. We discuss in this chapter the method of encoding HPFSMs by extending the traditional coding scheme that uses the MML principle. The coding scheme is in line with the two-part encoding of a PFSM model that we will discuss in the following sections. The first part encodes the HPFSM model and the second part encodes the data generated by the model. The two-part MML code length is the sum of the code lengths of the first part and the second part of the HPFSM model. From the artificially created HPFSM model, random strings or data tokens are generated and we compare the cost of encoding the HPFSM model for those tokens with the traditional PFSM model.

We then discuss the experiments performed on the UCI gathered, Activities of Daily Living (ADL) datasets. The dataset is a collection of daily activities performed by two individuals in their home settings that are captured through sensors. We discuss a method of encoding the dataset in the form of an HPFSM model. The initial models learnt are then inferred using Simulated Annealing and we do an analysis by using those models. The analysis is the prediction of individuals using the models. We do training on nearly half of the datasets and use the other half for testing purpose. Here we use accuracy of prediction as the evaluating criteria for the models.

The model two-part code length is compared with the traditional PFSM model and also with the one-state model. The results indicate that the HPFSM model gives the best compression when compared to the other models.

The chapter is organized to cover the contributions in the following way in different sections. Section 18.2 covers the idea of Probabilistic Finite State Machines, Section 18.3 covers MML encoding and Inference of PFSMs, and Section 18.4 covers HPFM models. The experiments on artificial datasets are shown in Section 18.5, and we finally conclude our contributions in Section 18.6.

18.2 Finite State Machines (FSMs) and PFSMs

Finite State Machine (FSM) is a mathematical model of computation which can effectively describe grammar of a language. The grammar described by this model of computation is called a regular grammar. The model plays an important role in several applied areas of computer science, of which, text processing, compiler and hardware design are the few common ones [20]. Finite State Machines can effectively represent regularities and patterns. For this reason, the words or tokens which are generated out of a natural language corpus can be effectively modelled using a Finite State Machine as the corpus contains many word repetitions.

A hypothetical grammar (FSM) generating or representing a collection of strings (sentences) in a finite alphabet might contain regularities that are not fully captured by the formal grammar. We can extend the simple FSM model to include some probabilistic structure in the grammar. The model is now termed as a Probabilistic Finite State Machine (PFSM). The grammar represented by a PFSM is called a probabilistic regular grammar. A PFSM, along with generating a set of possible strings, also defines a probability distribution on the set ([21], Section 7.1.2).

18.2.1 Mathematical Definition of a Finite State Machine

A Finite State Machine M is defined as a 5-tuple, where $M = <Q, \sum, \delta, q_0, F>$. The tuples are defined as below [11]:

- Q is the set of states of M
- $\sum$ is the set of input alphabet symbols
- δ is the transition function mapping $Q \times \sum$ to Q
- q_0 is the initial state
- F is the set of final states of M

18.2.2 Representation of an FSM in a State Diagram

The mathematical definition of a finite state machine can be easily transformed into a state diagram as shown in Figure 18.1. The FSM in Figure 18.1 corresponds to language L, where L = {CAB, CAAAB, BBAAB, CAAB, BBAB, BBB, CB}. The language is adopted from the reference [7].

All the strings in language L get accepted by the Finite State Machine in Figure 18.1. The FSM starts with state 0 and reads the string belonging to language L in a sequence by making state transitions. The final states shown in the FSM of Figure 18.1 are marked by double circles over the state numbers. After reading the complete string, the FSM again begins with state 0 to read the next string. Such FSM where the common prefixes of the different strings are marked by common states is known as the Prefix Tree Acceptor (PTA) of the strings of language L [19].

Let us again considering the language L, where L = {CAB, CAAAB, BBAAB, CAAB, BBAB, BBB, CB}. The language L is considered as data or observations and the strings in the data can be considered as tokens or words or even sentences. Let us assume that these tokens of language L are

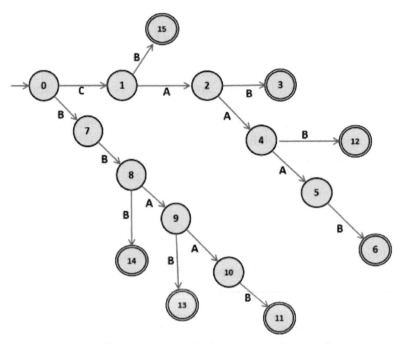

Figure 18.1   Finite State Machine accepting language L.

getting generated from an FSM whose true structure is not known. The tokens belonging to language L arrive in the same order as they are mentioned in the sequence and we separate them by a delimiter symbol # to distinguish the tokens from each other. The same language L now looks like L = {CAB#CAAAB#BBAAB#CAAB#BBAB#BBB#CB#}.

To model this sequence, we again start with some initial FSM and the initial FSM is the Prefix Tree Acceptor (PTA) of the language L. While reading the tokens one by one, whenever the symbol # is read, the machine again starts with the initial state of the FSM. In other words, the symbol # on the current state forces the machine to do a transition to the initial state from the current state. Figure 18.2 shows a hypothetical Finite State Machine generating the tokens of language L.

The FSM in Figure 18.2 lacks the expressibility of denoting regularities in terms of frequency of probability of occurrence of certain structures. Therefore, the finite state machines are incapable of representing regularities that are not fully captured by formal grammar. For this reason, the FSM

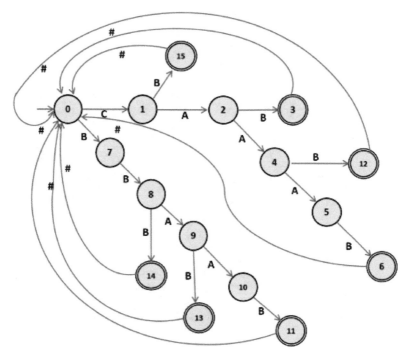

Figure 18.2   Hypothetical Finite State Machine generating L.

model is extended to include some probabilistic structure in the grammar. The extended model is now termed a Probabilistic Finite State Machine (PFSM).

In the FSM of Figure 18.2, the transition arcs are additionally labelled with transition probabilities to transform the representation into a PFSM representation ([21], Section 7.1.2).

Thus, for the same language L, where L = {CAB#CAAAB#BBAAB# CAAB#BBAB#BBB#CB#}, the PFSM is shown in Figure 18.3.

When we construct an FSM from a set of sample strings, we can estimate the transition probabilities by keeping a count of the number of times each arc of the graph is traversed [19]. The counts can be converted into probability estimates by dividing each count by the total count of all the arcs from that state. This turns the FSM to PFSM. The PFSM essentially tells the probability of transition from one state to another state on seeing a particular symbol from the finite alphabet set.

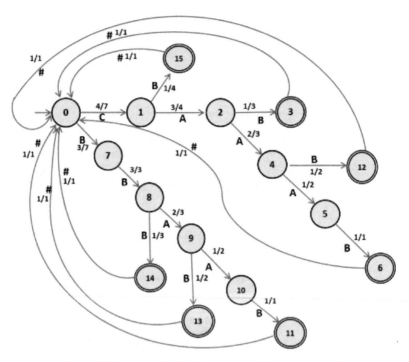

Figure 18.3   Probabilistic finite state machine generating L.

18.3 MML Encoding and Inference of PFSMs

MML works on the Bayesian principles of calculating the posterior probability of a hypothesis given the data and an elementary information theory concept based on Huffman code, converts the probability values into the form of code length in bits [21].

In probability theory and statistics, Bayes's theorem is a result of conditional probability, stating that for two events A and B, the conditional probability of A given B is the conditional probability of B given A is scaled by the relative probability of A compared to B [18].

Bayes's theorem is stated mathematically as the following equation:

$$Pr(A|B) = \frac{Pr(A\&B)}{Pr(B)} = \frac{Pr(B/A)Pr(A)}{Pr(B)} \qquad (18.1)$$

where A and B are events and $Pr(B) \neq 0$

- Pr(A) and Pr(B) are the probabilities of observing events A and B without regard to each other.

- Pr(A & B) is the probability of observing both events A and B.
- Pr(A|B), a conditional probability, is the probability of event A knowing that event B is true.
- Pr(B|A) is the probability of observing event B given that A is true.

MML has emerged as a powerful tool, not only in providing an encoding mechanism to find the code length of a PFSM but, it also plays an important role in the inductive inference of discrete structures such as PFSMs. This chapter discusses how an inductive hypothesis (PFSM) can be modelled from a finite set of sentences drawn from a finite alphabet set. An inductive hypothesis is said to be an abstraction over a set of sentences, but the inference of hypothesis tells how probable the model is in generating those sentences. So, we also discuss the method of inference using MML in this section.

18.3.1 Modelling a PFSM

Abstraction in the form of a PFSM is modelled using Minimum Message Length (MML). MML principle has its roots in the classical Bayesian theory of calculating the posterior probability of a hypothesis H. Computing the posterior probability equivalently means computing the product of the prior probability of the hypothesis and the probability of the data D generated in light of the hypothesis [2, 12]. This equivalently means to compute the sum of the following code lengths:

- Code length of the hypothesis H.
- Code length of the data D encoded using this hypothesis.

The problem of modelling an inductive hypothesis from a given set of observations becomes one of choosing between the competing models. Georgeff and Wallace [18] proposed the MML principle to help make the decision. The hypothesis that minimizes the sum of the above code lengths is regarded as the best one. Quantitatively, the sum can be put down as the following formula that calculates the two-part code length of a hypothesis in bits.

$$CodeLength(H|D) = CodeLength(H) + CodeLength(D|H)$$
$$= -\log_2 Pr(H) - \log_2 Pr(D|H) \quad (18.2)$$

18.3.1.1 Assertion code for hypothesis H

If the hypothesis is stated in the form of a PFSM, then the number of bits derived to encode it can be calculated. Let us consider the PFSM of Figure 18.2, which is the Prefix Tree Acceptor (PTA) of the

Gaines [7] data. Let us now call the collection as the observed data D. So, D = {CAB#CAAAB#BBAAB#CAAB#BBAB#BBB#CB#}. The definitions used in the context of finding the code length of a PFSM are enumerated below.

(1) S is the number of states in the FSM.
(2) Σ is the input alphabet set.
(3) V is the cardinality of the input alphabet set.
(4) n_{ik} is the number of transitions from state i on symbol k, where $k \in \Sigma$.
(5) M is the total number of arcs from all the states.
(6) a_i is the number of arcs leaving the current state i.
(7) t_i is the total number of times the state has already been left.

Note that $\sum_{i=1}^{S} a_i = M$.

The hypothesis H is encoded considering the number of states in the PFSM, number of arcs leaving each state of the PFSM, the labels on the arcs and the destinations. Wallace, 2005 (Section 7.1.3) [21] describes a better coding scheme and the coding scheme works as follows.

- The code begins by first stating the number of states in the PFSM. As the number of states is specified by S, it takes $\log^2$ S number of bits to state the number of states in the PFSM. For the PFSM in Figure 18.3, the number of states is 16 and the cost to state this is $\log_2$ 16 bits.
- Next the arcs leaving each state are stated. The number of possibilities for the number of arcs from each state is V, by considering a uniform distribution of the symbols from each state. Therefore, from each state, the arcs leaving each state are encoded in $\log_2$ V bits. The number of arcs leaving each state is quantified as a_i, where i is a state and $0 \le i \le (S - 1)$. The number of different possibilities for a_i is between 1 and V.
- The set of symbols leaving any state i is encoded next. The different number of ways of selecting a_i symbols from a set of V symbols is calculated as $\binom{V}{a_i}$.

  Therefore, the number of bits required to state this set of symbols from state i is, $\log_2 \binom{V}{a_i}$.
- The PFSM specification includes the destination state reached when a symbol is input on a current state i. The different a_i symbols on the current state i makes the PFSM transit on a_i states. As the destination belongs to one of the states from 0 to S − 1, the number of bits required to specify the destination states from current state i is $a_i \log_2$ S. But, for the arcs labelled #, the destination state is implied. Therefore, the number of bits required in this case is $(a_i - 1) \log_2$ S.

Using the above coding scheme any PFSM can be stated but, the code above contains some inefficiencies. The numbering of the states other than State 0 can be done in any arbitrary manner stating $(S - 1)!$ equal length code lengths for the same PFSM. Therefore, the redundancy is removed by obtaining a correction in the above code, by subtracting $(S - 1)!$ from the calculated code length of the PFSM. The code is still redundant as it permits descriptions of PFSMs some of whose states cannot be reached from the starting state. This amount of redundancy, however, is small and can be thus ignored ([21], Section 7.1.3).

For the PFSM of Figure 18.3, the asserted costs for the number of arcs, labels and destinations for each state are calculated and shown in Table 18.1.

Table 18.1 Code Length of PFSM from Figure 18.3 accepting D

State	a_i	Cost	Label(s)	Cost	Dest.(s)	Cost
0	2	$\log_2 V$	(C, B)	$\log_2 \binom{V}{2}$	(1,7)	$2\log_2 S$
1	2	$\log_2 V$	(A, B)	$\log_2 \binom{V}{2}$	(2,15)	$2\log_2 S$
2	2	$\log_2 V$	(B, A)	$\log_2 \binom{V}{2}$	(3,4)	$2\log_2 S$
3	1	$\log_2 V$	(#)	$\log_2 \binom{V}{1}$	(0)	0
4	2	$\log_2 V$	(A, B)	$\log_2 \binom{V}{2}$	(5,12)	$2\log_2 S$
5	1	$\log_2 V$	(B)	$\log_2 \binom{V}{1}$	(6)	$\log_2 S$
6	1	$\log_2 V$	(#)	$\log_2 \binom{V}{1}$	(0)	0
7	1	$\log_2 V$	(B)	$\log_2 \binom{V}{1}$	(8)	$\log_2 S$
8	2	$\log_2 V$	(A, B)	$\log_2 \binom{V}{2}$	(9,14)	$2\log_2 S$
9	2	$\log_2 V$	(A, B)	$\log_2 \binom{V}{2}$	(10,13)	$2\log_2 S$
10	1	$\log_2 V$	(B)	$\log_2 \binom{V}{1}$	(11)	$\log_2 S$
11	1	$\log_2 V$	(#)	$\log_2 \binom{V}{1}$	(0)	0
12	1	$\log_2 V$	(#)	$\log_2 \binom{V}{1}$	(0)	0
13	1	$\log_2 V$	(#)	$\log_2 \binom{V}{1}$	(0)	0
14	1	$\log_2 V$	(#)	$\log_2 \binom{V}{1}$	(0)	0
15	1	$\log_2 V$	(#)	$\log_2 \binom{V}{1}$	(0)	0

The cost columns in the table are added and the cost of encoding the hypothesis H without applying correction is shown by Equation (18.3).

$$\sum_{i=1}^{S} \log_2 \binom{V}{a_i} + \sum_{i=1}^{S} a_i \log_2 S + \sum_{i=1}^{S} \log_2 V + \log_2 S$$

$$= \sum_{i=1}^{S} \log_2 \binom{V}{a_i} + M \log_2 S + S \log_2 V + \log_2 S \qquad (18.3)$$

After applying the correction to the above code by subtracting $(S-1)!$ from the equation above, the code length for the hypothesis H for the PFSM in Figure 18.3 is given by Equation (18.4).

$$CodeLength(H) = \sum_{i=1}^{S} \log_2 \binom{V}{a_i} + M \log_2 S + S \log_2 V$$

$$+ \log_2 S - \log_2 (S-1)! \qquad (18.4)$$

The formula in Equation (18.4) is used further in calculating the first-part code length of any PFSM. The second part-code length, which encodes the data D, assuming the hypothesis to be true, is discussed in the section following this.

18.3.1.2 Assertion code for data D generated by hypothesis H

This part of the code length asserts the data D generated by the hypothesis H. From each state of the PFSM, there are multiple transitions on the arcs leaving the state; therefore, the distribution from each state appears to be multinomial.

The code length of the data D generated by the hypothesis H is given by Equation (18.5).

If it is desired to treat the probabilities as an essential feature of the inferred model, a small correction from ([21], Section 5.2.13) can be added to the above expression, as detailed in ([21], Section 7.1.6), given by approximately $(1/2 \, (\pi(a_i - 1)) - 0.4)$ for each state. For the state having only one exit arc, there is no need to calculate the transitional probability.

$$CodeLength(D|H) = \sum_{i=1}^{S} \log_2 \frac{(t_i + a_i - 1)!}{(a_i - 1)! \Pi_k(n_{ik}!)} \qquad (18.5)$$

The total two-part code length for the PFSM is written as CodeLength(H) + CodeLength(D|H), which is approximately equal to the following expression in Equation (18.6).

Total Two-Part CodeLength

$$
\begin{aligned}
&= \sum_{i=1}^{S} \left\{ \log_2 \frac{(t_i + a_i - 1)!}{(a_i - 1)! \Pi_k (n_{ik}!)} + \log_2 \binom{V}{a_i} \right\} \\
&\quad + M \log_2 S + S \log_2 V + \log_2 S - \log_2 (S - 1)!
\end{aligned}
\tag{18.6}
$$

18.3.2 Inference of PFSM Using MML

The number of PFSMs that can exist for a given data sequence is computationally intractable [17]. For S number of states in a PFSM, the number of PFSMs that all can account for the same data sequence, is exponential in S. According to Gold [9], Angluin [1] and Feldman [6], the problem of searching the minimal PFSM from a given data sequence is known to be NP-complete. Therefore in searching the best PFSM to account for a given data sequence along with making the problem of searching tractable, a trade-off between the guarantee of optimality of the solution is done with tractability considerations in mind, and this is achieved using heuristics.

To infer a stochastic PFSM with transition probabilities that best reflects a given data sequence, Raman [15], Raman and Patrick [14], Clelland [3], Clelland and Newlands [4], Raman and Patrick [14], Collins and Oliver [5], and Hingston [10], propose various approaches. The method on inferencing works along the lines of building a minimal or maximal canonical PFSM. The minimal PFSM to start with is the one state PFSM where the states are split in each iteration until no further split is possible. The maximal PFSM to begin with is the Prefix Tree Acceptor (PTA) of the data sequence. The states in the PTA are progressively merged until no further merge is possible.

In this section, we discuss two methods of inferencing a PFSM, represented as a PTA of the data sequence. The methods work along the lines of ordered merging and random merging of states in the PFSM. The method of ordered merging is completely novel and proposed by us. This method uses a greedy search heuristic to find a near optimal PFSM and most suitable for applications where optimality can be compromised for quickness of the solution. The second method proposed by Raman and Patrick [14] uses a simulated annealing heuristic to find an optimal solution. This method can guarantee an accurate solution by appropriate setting of the temperature and rate of cooling. We have re-implemented this method in our own way by doing minimal modifications in the original algorithm in the way that suits our needs. By optimal solution, we mean the minimum two-part code length PFSM and the objective function to get the optimal solution is the Minimum Message Length (MML) principle.

18.3.2.1 Inference of PFSM by ordered merging (OM)

For inference using Ordered Merging (OM), the induction process begins by considering the input in the form of a PTA of the data sequence. The node pairs are merged in stages satisfying two major constraints:

- First, the merging always remains deterministic. That is, when two states are merged, the transitions on any input symbol on the merged states should be unambiguous. Any symbol from the merged states should not lead transitions to more than one state. If this is not satisfied, the second condition is never tried.

- Second, the two-part code length of the new PFSM after merge is lesser than the two-part code length of the PFSM before merge. Since the method works along the lines of searching using greedy search heuristic, the merge is only tried when the two-part code length of the PFSM after merge is lesser than the two-part code length of the PFSM before merge.

The stages of the node pair merges are given in detail in the subsections below.

18.3.2.1.1 *First stage merge*

In the first stage of the merge process, the final states of the initial PFSM are merged. The final states are those states in the PFSM that have transitions on input symbol #. This transition on input symbol # leads back to the initial state of the PFSM.

To see the effect of this merging, we reconsider the PFSM in Figure 18.3, which is the PTA of the Gaines data. The two-part code length calculated for this PTA using Equation (18.6) is 181.205 bits. The final states with input symbol # transitions from them, are merged and the resulting PFSM after merging final states is shown in Figure 18.4. The two-part code length of the new PFSM is 136.365 bits.

An important observation to be noted here is, it is not necessary that all the final states get merged as one by applying the First Stage merge Process. So, there might be chances of still getting more than one final states after the First Stage Merge is over. This is attributed to the constraints put on merging.

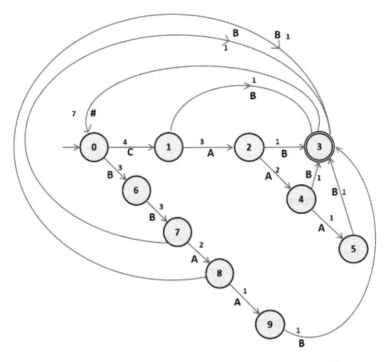

Figure 18.4 PFSM with final states merged from Figure 18.3.

18.3.2.1.2 *Second stage merge*

In the second stage of the merge process, the states directly connected to the final states are merged with each other. No merging with the final states is done here. The process of merging through the list of states directly connected with the final state continues until no further merging turns beneficial. The effect of applying the Second Stage Merge is seen in Figure 18.5. The two-part code length of this PFSM is 66.6234 bits.

18.3.2.1.3 *Third stage merge*

In the third stage of the merge process, pair of states are merged where one state is the final state and the other state is the state connected to it. We loop through the list until we reach a PFSM whose code length is the minimum. This PFSM can now be termed as the MML PFSM. This kind of stage-wise merging is more systematic as opposed to the random merge pairs in the Beam Search Algorithm [14]. Merging states in Third Stage of the inference process generates the same PFSM as in Figure 18.5.

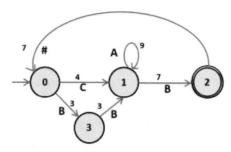

Figure 18.5 PFSM with second stage states merged from Figure 18.4.

18.3.2.1.4 *Ordered merging (OM) algorithm*

We formally write the method of inferencing through Ordered Merging in the form of an algorithm. The merges achieved in different stages are mentioned as three separate algorithms. Algorithm 2 is very much similar to Algorithm 1. The only difference between the two algorithms is, the merging is applied on different lists in both the cases. Algorithm-1 considers the list of final states for merging the states, whereas Algorithm-2 considers the list of states directly connected to final states for merging the states. The two-part code length calculated using MML is used as an objective function to obtain the minimum code length PFSM.

Algorithm 1 Ordered merging first stage

Input: PTA
Output: Reduced PFSM after First Stage Merge
1:   ListOfFinalStates $\leftarrow$ List of Final States
2:   NoOfFinalStates $\leftarrow$ Number of Final States
3:   old$_{MML}$ $\leftarrow$ CodeLengthOfCurrentPFSM
4:   for $i \leftarrow 1$ to NoOfFinalStates $- 1$ do
5:       for $j \leftarrow i + 1$ to NoOfFinalStates do
6:           if (CanBeMerged(ListOfFinalStates$[i]$,ListOfFinalStates$[j]$)) then
7:               new$_{MML}$ $\leftarrow$ Merge(ListOfFinalStates$[i]$,ListOfFinalStates$[j]$)
8:               if (new$_{MML}$ $\leq$ old$_{MML}$) then
9:                   old$_{MML}$ = new$_{MML}$
10:                  $i \leftarrow i - 1$
11:                  break
12:              else
13:                  UndoMerge()
14:              end if
15:          end if
16:      end for
17: end for

Algorithm 2 Ordered merging second stage

Input: PFSM resulting from First Stage Merge
Output: Reduced PFSM after Second Stage Merge
1:   ListOfStates ← List of states directly connected to Final States
2:   NoOfStates ← Number of states directly connected to Final States
3:   old_{MML} ← CodeLengthOfCurrentPFSM
4:   for i ← 1 to NoOfStates − 1 do
5:       for j ← $i + 1$ to NoOfStates do
6:           if (CanBeMerged(ListOfStates$[i]$,ListOfStates$[j]$)) then
7:               new_{MML} ← Merge(ListOfStates$[i]$,ListOfStates$[j]$)
8:               if ($new_{MML} \leq old_{MML}$) then
9:                   old_{MML} = new_{MML}
10:                  i ← $i − 1$
11:                  break
12:              else
13:                  UndoMerge()
14:              end if
15:          end if
16:      end for
17: end for

18.3.2.2 Inference of PFSM using simulated annealing (SA)

We are basing the inference of a PFSM using Simulated Annealing (SA) by following a path that considers random merging of nodes instead of ordered merging. Raman and Patrick [14] use the method of inference using SA to get the minimal code length PFSM and we briefly explain the method in this section.

To understand the method of inferencing using SA, we first try to explain what Annealing is. Annealing is a physical process in which metals are heated to very high temperature and then they are gradually cooled. The high temperature causes the electrons of the metal to emit photons and in this process of doing so, the metal gradually descends from high energy state to low energy state. The emitted photons may bump into another electron, causing it to move to high energy state, but the probability of happening this decreases with cooling. The probability p is quantified as $p = e^{-\frac{\Delta E}{kT}}$, where ΔE is the positive change in energy level of electron, T is temperature and k is Boltzmann's constant. The rate of cooling is called annealing schedule and it plays a very important role in the formation of the final product. Rapid cooling will prohibit the electrons to descend to lower energy state and, as a result, there will be formation of regions of stable high energy. Too much slow cooling will be a waste of time. Therefore, there has to

---

Algorithm 3 Ordered merging third stage

Input: PFSM resulting from Second Stage Merge
Output: Minimum Code Length PFSM or the Inferred PFSM
1:   $\text{old}_{MML} \leftarrow \text{CodeLengthOfCurrentPFSM}$
2:   $\text{new}_{MML} \leftarrow 0$
3:   $\text{ListOfFinalStates} \leftarrow \text{List of Final States}$
4:   $\text{NoOfIterations} \leftarrow \text{Initial Number Of Iterations}$
5:   $i \leftarrow 0$
6:   while ($\text{new}_{MML} < \text{old}_{MML}$ OR $i \leq \text{NoOfIterations}$) do
7:       $\text{state1} \leftarrow \text{Random State from ListOfFinalStates}$
8:       $\text{state2} \leftarrow \text{Random State from States connected to state1}$
9:       if CanBeMerged(state1,state2) then
10:          $\text{new}_{MML} \leftarrow \text{Merge(state1,state2)}$
11:          if ($\text{new}_{MML} \leq \text{old}_{MML}$) then
12:              $\text{old}_{MML} = \text{new}_{MML}$
13:              $\text{new}_{MML} \leftarrow 0$
14:          else
15:              UndoMerge()
16:              $i \leftarrow i + 1$
17:          end if
18:      else
19:          $i \leftarrow i + 1$
20:      end if
21: end while

---

be an optimum choice of the annealing schedule and that is determined empirically.

18.3.2.2.1 *Simulated annealing (SA)*

In simulated annealing procedure, Boltzmann's constant is irrelevant as it is specific to the physical process. Therefore, the probability formula now becomes $p = e^{-\frac{\Delta E}{kT}}$. As ΔE refers to the change in the energy state of an electron in Annealing procedure, in Simulated Annealing it refers to the change in objective function. Thus, in this case, ΔE is the change in the two-part code length of the current PFSM. The temperature in annealing is set in Kelvin and in SA, the temperature is set to some value which is number of bits. The value is again determined empirically as in annealing.

The modified probability is now given by

$$p' = e^{\frac{-(new_{CodeLength}-old_{CodeLength})}{T}},$$

where $\text{new}_{CodeLength}$ is the code length of the changed PFSM after applying node merge and $\text{old}_{CodeLength}$ is the code length of the PFSM before applying merge. The change in the objective function is guaranteed to be a positive quantity as the probabilistic acceptance is only tried when the new$_{\text{CodeLength}}$ is larger than the old$_{\text{CodeLength}}$. For a negative change in the objective function, where the new$_{\text{CodeLength}}$ is always smaller than the old$_{\text{CodeLength}}$, the merge is always considered as it results in better solution.

The Simulated Annealing method of inferencing is formalized as an algorithm in Section 18.3.2.2.2.

18.3.2.2.2 *Simulated annealing (SA) algorithm*

Algorithm 4 Simulated annealing

1:   old$_{\text{MML}}$ ← Code Length of Current PFSM
2:   Temperature ← Initial Temperature
3:   CurrentState ← Initial State of current PFSM
4:   while Temperature = 0 do
5:       RandomState1 ← Random state of current PFSM
6:       RandomState2 ← Random state of current PFSM
7:       while RandomState1 = RandomState2 do
8:          RandomState1 ← Random state of current PFSM
9:          RandomState2 ← Random state of current PFSM
10:     end while
11:     if CanBeMerged(RandomState1,RandomState2) then
12:       new$_{\text{MML}}$ ← Merge(RandomState1,RandomState2)
13:       if new$_{\text{MML}}$ ≤ old$_{\text{MML}}$ then
14:          old$_{\text{MML}}$ = new$_{\text{MML}}$
15:       else
16:          $p = e^{\frac{-(\text{newMML}-\text{oldMML})}{\text{Temperature}}}$
17:          RandomNumber = Random(0, 1)
18:          if RandomNumber ≤ p then
19:             old$_{\text{MML}}$ = new$_{\text{MML}}$
20:          end if
21:       end if
22:     end if
23:     Temperature ← Temperature − 1
24: end while

18.4 Hierarchical Probabilistic Finite State Machine (HPFSM)

The research work done earlier that encodes a hypothesized PFSM, considers the PFSM to be structurally non-hierarchical. Various induction methods using MML have been proposed and they all aim at resulting in a minimal PFSM, which is again structurally non-hierarchical. But, a careful observation at the data observations would reveal that, some of the subsets can be generated by small internal PFSMs without the need to get generated out of the big single PFSM (non-hierarchical). And the small PFSMs are tied together in the outer structure resulting in some kind of hierarchy.

This viewpoint is extremely essential to understand the things happening in ground reality. If we consider an example, where we have a set of observations and the observations are the words belonging to different languages. In those sets of observations, if we try constructing a PFSM that best describes those sets of observations, the code length would be enormous as it has to include the vocabularies of all the languages. Also, if the observations result in heavy use of one language than the other, then there are infrequent transitions from one language to the other language. So, constructing a single non-hierarchical PFSM would turn out to be more expensive in this case. We can do this in a cheaper way by constructing a hierarchical PFSM for the set of observations. For enormous enumerations of the data sequence, the cost of encoding the hypothesis H and the data D generated by the hypothesis would be cheaper than the cost of encoding a non-hierarchical PFSM. But, on the other hand, if we have very less enumerations (which is quite improbable), we would be unnecessarily paying for the complex hierarchical structure.

Another example why coding in hierarchy would turn out to be beneficial could be understood this way. Let us say we have a big machine that represents our movements in a day, in the form of state transitions, from one place to another. Broadly speaking, the places that we may visit are, "university or work place", "home" and "city". The machine representing this situation is seen in Figure 18.6. We start from a local place say home. Home represents one small internal PFSM. At "home", we do state transitions locally more often by visiting different places at "home". There is one point of entry (start state) to "home" and the same point can be used as the exit point from this "home" local machine. Then, we transit from "home" to another internal PFSM, say "university". We enter into the start state of the "university" local PFSM and again we perform state transitions by moving around different places in the "university". The start state of the "university" internal PFSM is also the exit state from it, as we had in the "home" internal PFSM. If we are to specify a record of the daily movements that involves moving around various places, in the form of code length in bits, then considering the whole picture as a non-hierarchical PFSM and thereafter calculating the code length would turn to be very expensive. Whereas, if we do the same through hierarchical coding mechanism, we would definitely get a cheaper encoding of the daily movements of the person.

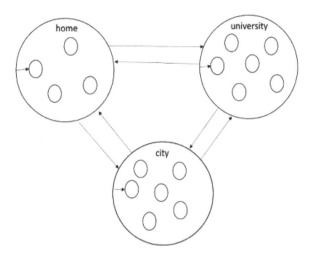

Figure 18.6 An example of hierarchical PFSM.

18.4.1 Defining an HPFSM

A hierarchical PFSM consists of an outer PFSM whose states can internally contain inner PFSMs (or inner HPSMs). We refer to the hierarchical PFSM as HPFSM to keep it short. An example of such an HPFSM is shown in Figure 18.7. The probability of transition on a particular symbol from any state in the HPFSM of Figure 18.7 is not shown, but from every state the transitions on the different symbols are seen equally probable. That, a multinomial distribution with a uniform prior is considered from each state on all symbols.

The behavior of a simple PFSM model can be understood in a very concise manner in terms of an HPFSM and the obvious benefit is a less two-part code length machine that still represents the same grammar. We refer to the HPFSM model in Figure 18.7, which is a special case, where there are three outer states and each of the outer PFSMs internally contains PFSMs with three states inside, but in general, we can have as many outer and inner states as we may like in the HPFSM model. The structure is explained like this. The HPFSM has three states in the outer structure labelled S1, S2 and S3. The outer states internally contain PFSMs with three states in each PFSM. The outer PFSM has an initial state S1 and this is the string generation point in the HPFSM. Each internally contained PFSM has a starting state and the same state is used as an exit state to transit to other PFSMs in the outer structure. Each internally contained PFSM can independently generate strings

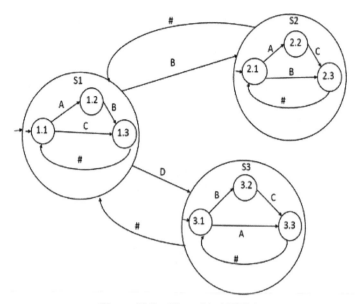

Figure 18.7   Hierarchical PFSM.

separated by the delimiter symbol # or it can communicate with the other PFSMs through the outer transitions and can generate strings that include the outer symbols. Like, for example, the kind of strings produced by the HPFSM of Figure 18.7 can look like D = {AB#BAC##C#DBC#BC#A##...}. Each internally contained PFSM has its own input alphabet symbols and similarly the outer structure also has one.

18.4.2 MML Assertion Code for the Hypothesis H of HPFSM

Before starting with the two-part code length calculation for the HPFSM, we refer to the following definitions:

- S_{outer} is the number of states in the outer PFSM.
- V_{outer} is the cardinality of the input alphabet set in the outer PFSM including the delimiter symbol #.
- The states in the outer PFSM are labelled as $S1, S2, \ldots, S_{S_{outer}}$.
- The number of states in the inner PFSMs are denoted as $S_{internal1}$, $S_{internal2}, \ldots, S_{internalS_{outer}}$.
- The cardinalities of the input alphabets in the inner PFSMs are denoted as $V_{internal1}, V_{internal2}, \ldots, V_{internalS_{outer}}$.

- The number of arcs leaving any state in internal PFSM Sj is denoted by a_{sji}, where $1 \leq j \leq S_{outer}$ and $1 \leq i \leq S_{internalj}$.
- The number of transitions on symbol k from any current state i in any internal PFSM S_j is denoted as n_{jik}, where k can be a symbol from the input alphabet set of S_j or the input alphabet set of outer PFSM.

The coding scheme described below is explained in reference to S1 internal PFSM. The other PFSMs are encoded in a similar fashion.

1. The code begins with the number of states in internal PFSM, which is $S_{internal1}$, and the code length is calculated as $\log_2 S_{internal1}$ bits.
2. For each state in the inner PFSM S1, the number of arcs leaving the state are coded. The number of arcs leaving any state depends on the cardinality of the input alphabet set. Since state 1 of internal PFSM S1 has outgoing arcs to internal PFSM S1 and also outgoing arcs to other states in the outer structure; therefore, a selection from 1 to $V_{outer} + V_{internal1}$ is made giving a code length of $\log_2 (V_{outer} + V_{internal1})$ bits. We calculate the combined cardinality denoted by the expression $V_{outer} + V_{internal1}$ by considering unique input alphabets in the two alphabet sets. So, if # appears in the input alphabet set of the both the internal and the outer structure, it is counted as 1 and likewise any other character other than #.
3. For each state, the labels on the arcs are coded. This is done by making a selection of a_{sj1} symbols from the set of ($V_{outer} + V_{internal1}$) symbols giving a code length of $\log_2 (V_{outer} + V_{internal1})\, a_{sj1}$ bits, if the state is the starting state of the internal PFSM. Otherwise for the other states, the code length is $\log_2 V_{internal1}$ bits. Here, $1 \leq j \leq S_{outer}$.
4. The code for the destination states is calculated from each state. From the starting state, the destination can be any of the internal states of that PFSM or it can be of the states in outer structure. In the outer structure, the destination is always the initial state of other internal PFSMs. Therefore, for the starting state, the code length encoding the destinations is $a_{sj1} \log_2 (S_{outer} + S_{internal1})$ bits. For states other than the starting state, the coding can be done in $\log_2 S_{outer}$ bits. For the arcs labelled with # delimiter symbol the destination is already known and the coding can be done in $(a_{sj1} - 1) \log_2 S_{internal1}$ bits.

The coding scheme above generates Table 18.2, which shows encoding of the first part code length of the S1 internal PFSM. The other internal PFSMs are encoded similarly and are shown in Tables 18.3 and 18.4.

Table 18.2   Code Length of internal PFSM S1 from Figure 18.7

State	$a_{s_{j1}}$	Cost	Label(s)	Cost	Dest.(s)	Cost
1.1	4	$\log_2\left(V_{outer}+V_{internal1}\right)$	(A,B,C,D)	$\log_2\binom{V_{outer}+V_{internal1}}{4}$	(1.2, 1.3, S2, S3)	$4\log_2\left(S_{outer}+S_{internal1}\right)$
1.2	1	$\log_2 V_{internal1}$	(B)	$\log_2\binom{V_{internal1}}{1}$	(1.3)	$\log_2 S_{internal1}$
1.3	1	$\log_2 V_{internal1}$	$(\#)$	$\log_2\binom{V_{internal1}}{1}$	(1.1)	0

Table 18.3   Code Length of internal PFSM S2 from Figure 18.7

State	$a_{s_{j2}}$	Cost	Label(s)	Cost	Dest.(s)	Cost
2.1	3	$\log_2\left(V_{outer}+V_{internal2}\right)$	$(A,B,\#)$	$\log_2\binom{V_{outer}+V_{internal2}}{3}$	(2.2, 2.3, S1)	$2\log_2\left(S_{outer}+S_{internal2}\right)$
2.2	1	$\log_2 V_{internal2}$	(C)	$\log_2\binom{V_{internal2}}{1}$	(2.3)	$\log_2 S_{internal2}$
2.3	1	$\log_2 V_{internal2}$	$(\#)$	$\log_2\binom{V_{internal2}}{1}$	(2.1)	0

Table 18.4   Code Length of internal PFSM S3 from Figure 18.7

State	$a_{s_{j3}}$	Cost	Label(s)	Cost	Dest.(s)	Cost
3.1	3	$\log_2\left(V_{outer}\right)+V_{internal3}$	$(B,A,\#)$	$\log_2\binom{V_{outer}+V_{internal3}}{3}$	(3.2, 3.3, S1)	$2\log_2\left(S_{outer}+S_{internal3}\right)$
3.2	1	$\log_2 V_{internal3}$	(C)	$\log_2\binom{V_{internal3}}{1}$	(3.3)	$\log_2 S_{internal3}$
3.3	1	$\log_2 V_{internal3}$	$(\#)$	$\log_2\binom{V_{internal3}}{1}$	(3.1)	0

The final equation that encodes the first part code length of the HPFSM is given in Equation (18.7). The start state of each internally contained PFSM is handled separately in the equation as it has transitions to outer states and that is the reason that variable i starts with 2 in the Equation (18.7).

$$
\begin{aligned}
CodeLength(H) = \sum_{j=1}^{S_{outer}} & \left(\sum_{i=2}^{S_{internal_j}} \log_2 \binom{V_{internal_j}}{a_{s_{ji}}} \right. \\
& + \log_2 \binom{V_{outer} + V_{internal_j}}{a_{s_{ji}}} \\
& + (S_{internal_j} - 1) \log_2 V_{internal_j} \\
& + \log_2(V_{outer} + V_{internal}) + \sum_{i=2}^{S_{internal_j}} a_{s_{ji}} \log_2 S_{internal_j} \\
& \left. + a_{s_{ji}} \log_2(S_{outer} + S_{internal_j}) + \log_2 S_{internal_j} \right) \\
& + \log_2 S_{outer}
\end{aligned}
\tag{18.7}
$$

18.4.3 Encoding the transitions of HPFSM

Again, assuming a uniform prior over a multinomial distribution case, the probability of transition on any symbol k from current state i in any internal PFSM S_j, is denoted as $\frac{(n_{ji_k}+1)}{(n_{ji}+a_{n_{ji}})}$, where n_{jik} represents the number of transitions from current state i in current internal PFSM S_j on symbol k and n_{ji} represents the total number of transitions on all symbols from the current state. The number of bits required to code the transitions on symbol k is negative logarithm of the transition probability. If we sum over the probabilities for all the symbols from the current state, then the total number of bits required to encode the transitions is given by the following equation:

$$
\frac{(n_{ji} + a_{s_{ji}})!}{\Pi_k(n_{ji_k})!}
\tag{18.8}
$$

where $1 \leq j \leq S_{outer}$, $1 \leq i \leq S_{internalj}$ and k is a symbol from the input alphabet sets of $S_{internalj}$ and S_{outer}.

Equation (18.9) calculates the second part code length of the HPFSM in Figure 18.7.

$$CodeLength(D|H) = \sum_{j=1}^{S_{outer}} \left(\sum_{i=1}^{S_{internal_j}} \log_2 \frac{(n_{ji} + a_{s_{ji}})!}{\Pi_k n_{ji_k}!} \right) \qquad (18.9)$$

The complete two-part code length for the HPFSM is calculated by adding Equations (18.7) and (18.9).

18.5 Experiments

The experiments are performed on the artificial datasets generated by the HPFSM of Figure 18.7 and on the UCI gathered Activities of Daily Living (ADL) datasets. In both the experimental situations, we compare the cost of the hierarchical model with the non-hierarchical model and the one-state model. In the experiments performed with the ADL datasets, we first describe a method of learning the initial HPFSM models from the datasets of the individuals. This learning is followed by induction using the Simulated Annealing (SA) search to obtain the optimal HPFSM models. We then finally perform a prediction of individuals based on the sequence of movements from the HPFSM models learnt.

18.5.1 Experiments on Artificial datasets

18.5.1.1 Example-1

We use the HPFSM of Figure 18.7 to generate random data strings or tokens of variable string lengths. As a recap, the HPFSM in Figure 18.7 has three outer states and each of the three outer states has an internal vocabulary of four characters including the delimiter symbol #. The outer PFSM has an input alphabet size of three characters. The strings are generated by setting up initial transition probabilities on all the transition arcs of the HPFSM model. The probabilities for this data-generating process have been set, in this first example, to be uniform in nature. For example, let us consider the state 1.1 in the HPFSM model. State 1.1 has 4 transitions to other states. That is, on characters A and C, states 1.2 and 1.3 are reached respectively from state 1.1 in the same internal PFSM S1. On characters B and D, outer states S2 and S3 are reached from state 1.1. If all the possibilities are considered equally probable in the process of string generation, then all the transition arcs are initially set to probabilities of 0.5 from state 1.1. Similarly, the probabilities on other transition arcs are set in the HPFSM model and strings get generated eventually. This constitutes an artificial dataset that is hierarchical as it is generated from the hierarchical structure. The smallest number of strings that we generate are 5 and going up to a maximum of 5000 strings. The string length of the strings that get generated from the HPFSM model are usually 1 or 2. This is because, each internal PFSM has three states internally and the final states in the internal PFSMs either read single character from the start state or 2 characters from the start state of the internal PFSMs.

Figures 18.8–18.10 show a comparison of the two-part code lengths computed using MML between the various models. We do a comparison of the HPFSM model with the initial non-hierarchical PFSM that is simply shown as PFSM (PTA) in the figures. The PFSM (PTA) models are inferred and what we obtain is the inferred PFSM model. The inferred PFSM model is results in shorter two-part code length model than the PFSM (PTA) model, as expected. We also compare with the one-state model.

Figure 18.8 shows the comparison on number of random strings from 2 in number to 80 strings. For the number of random strings from 2 to 20 in Figure 18.8, the one-state PFSM model gives the best compression when compared to the other models. The HPFSM model on the other hand results in the highest two-part code length for such less number of strings. This is obvious as the structure encoding is expensive for such a small number of strings. But as the number of strings further increases, the HPFSM model starts showing the least two-part code length when compared to other models.

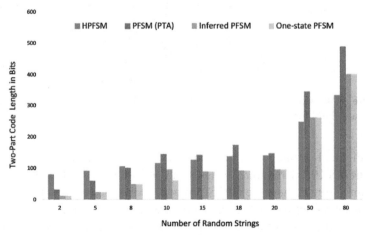

Figure 18.8 Code Length comparison between HPFSM, PFSM (PTA), inferred PFSM and one-state PFSM for random string 2–80.

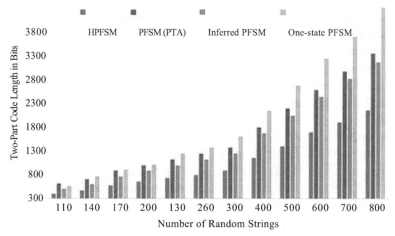

Figure 18.9   Code Length comparison between HPFSM, PFSM (PTA), inferred PFSM and one-state PFSM for random strings 110–800.

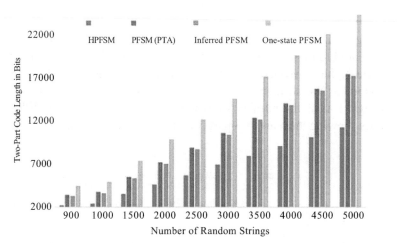

Figure 18.10   Code Length comparison between HPFSM, PFSM (PTA), inferred PFSM and one-state PFSM for random strings 900–5000.

Figure 18.9 shows the comparison on random strings from 110–800 and Figure 18.10 on random strings from 900–5000. The two figures show a common trend where the HPFSM model shows the best compression and the one-state model shows the worst compression for the dataset comprising the random strings in all the cases. The amount of compression achieved by the HPFSM model for 5000 number of random strings is 28.54% more than the

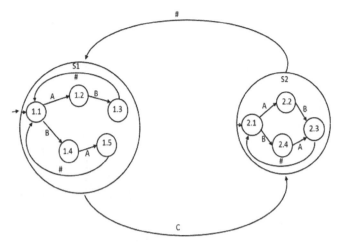

Figure 18.11 Hierarchical PFSM.

inferred PFSM model. Whereas the same HPFSM model shows 43.57% more compression than the one-state model for 5000 number of random strings. We calculate these percentages by calculating the difference in the two-part code length code length of the HPFSM model with the other models in comparison. The difference in code lengths is then divided by the two-part code length of the compared model and shown as percentage.

18.5.1.2 Example-2

We consider another example of an HPFSM as shown in Figure 18.11. In this example, the outer structure of the HPFSM model has two outer states and the two outer states internally contain 5 and 4 states, respectively, in their inner PFSMs. The internal vocabularies of the two internal PFSMs are the same. This time, instead of setting uniform prior probabilities in the transition arcs, we set non uniform probabilities. The reason for doing so comes from the motivation that made us think about the idea of HPFSMs. We set to high, the probabilities of transition in the internal arcs and likewise, in the outer transitions the probability values are set to low value. This is because the internal transitions are more frequent than the outer transitions. Variable number of random strings get generated out of the HPFSM model in Figure 18.11 and we do a similar analysis of the model two-part code length for the random strings generated. The two-part code length of the HPFSM model is then compared against the inferred PFSM and the one-state models.

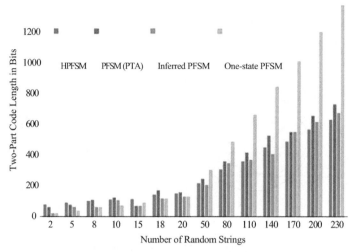

Figure 18.12   Code Length comparison between HPFSM, PFSM (PTA), inferred PFSM and one-state PFSM models for random strings 2–230 for the PFSM of Figure 18.11.

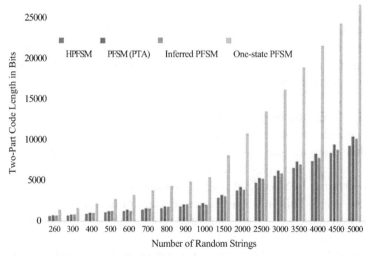

Figure 18.13   Code Length comparison between HPFSM, PFSM (PTA), inferred PFSM and one-state PFSM models for random strings 260–5000 for the PFSM of Figure 18.11.

Figures 18.12 and 18.13 show the comparison of the two-part code lengths of the HPFSM model with other models.

In Figure 18.12, we show a comparison on the number of random strings from 2 to 230. The HPFSM model shows the largest two-part code length for the number of random strings less than 8. This is understandable as the model is fairly complex in structure for such small number of strings. The one-state model on the other hand shows the least two-part code length for such small number of random strings. But as the number of strings further increase, the trend completely reverses. The HPFSM model gives the least two-part code length and the one-state model results in the largest two-part code length model.

In Figure 18.13, the comparison is shown on the number of random strings from 260 to 5000 and a similar observation follows. The two-part code length in bits obtained from the HPFSM model for 5000 number of random strings is 8728.45 bits and for the inferred PFSM model, the two-part code length is 9534.76 bits. The one-state model on the other hand results in 25421.8 bits for the same dataset. Therefore, for 5000 number of random strings, the probability of the HPFSM model of having generated those strings as compared to the inferred PFSM model is quantified as $\frac{29534.76}{(29534.76+28728.45)}$. Also, the HPFSM model is $\frac{225421.8}{(225421.8+28728.45)}$ more probable than the one-state model in generating the dataset of 5000 strings.

18.5.2 Experiments on ADL Datasets

The results are also computed using the Activities of Daily Living (ADL) datasets gathered from the UCI Machine Learning Repository (see Table 18.5). A short version of the dataset is presented in Table 18.5.

This dataset comprises information regarding the ADL performed by the two users on daily basis in their own home settings and summing up to 35 days of fully labelled data [13]. We call the individuals "Person-A" and "Person-B". Each individual dataset gives a description of the start time and end time of the event, the location of the event captured using sensors and the place where it happened. The five different places (bathroom, kitchen, bedroom, living and entrance) are initially considered as five outer states in the HPFSM, each of which has its own internal PFSM. The sequence

Table 18.5 Person-A activities of daily living (ADL)

Start Time	End Time	Location	Type	Place
28/11/2011 2:27	28/11/2011 10:18	Bed	Pressure	Bedroom
28/11/2011 10:21	28/11/2011 10:21	Cabinet	Magnetic	Bathroom
.	.	.	.	.
.	.	.	.	.
.				
12/12/2011 0:31	12/12/2011 7:22	Bed	Pressure	Bedroom

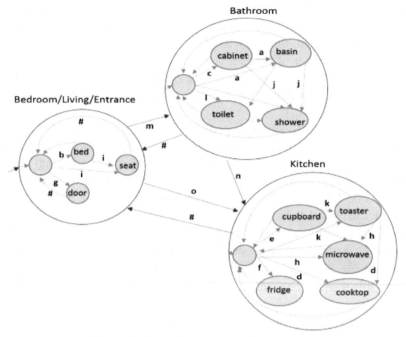

Figure 18.14   Inferred HPFSM for Person-A.

of transitions is captured as strings and encoded using the HPFSM. The conversion of the sequence of transitions into strings is done by assigning unique symbols to distinct locations in Table 18.5 and the change of place in the table is noted down as the end of sentence. This change of place inserts a delimiter symbol into the sequence formed so far. The HPFSM is inferred and we show as an example in Figure 18.14, the inferred HPFSM model for "Person-A" with three outer states. Three outer states (bedroom, living and entrance) get merged, resulting, finally in three outer states, as seen in the HPFSM of Figure 18.14. We similarly learn the HPFSM model for "Person-B" and the inferred HPFSM model for "Person-B" is shown in Figure 18.15.

Table 18.6 compares the two-part code length of the HPFSM and the non-hierarchical PFSM. The table reveals that, if the tokens generated by the activities of individuals, are encoded using HPFSM, the cost of encoding is less. This is evident from the two-part code length of the HPFSM and simple PFSM (or simply PFSM or non-hierarchical PFSM). We also compare with the one-state PFSM code length.

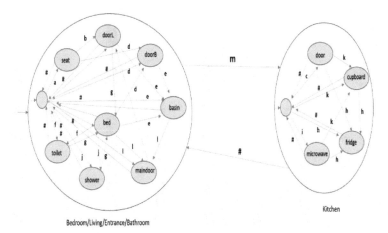

Figure 18.15 Inferred HPFSM for Person-B.

Table 18.6 Code length in bits given by various models for the ADL of individuals

	Initial HPFSM	Inferred HPFSM	Initial PFSM (PTA)	Inferred PFSM	1-state PFSM
Person-A	1235.10	1095.70	1691.02	1143.39	1143.39
Person-B	3062.52	3022.64	4098.80	3374.84	4036.83

A more useful analysis of the HPFSM models learnt from the training datasets is done and that is, prediction of the individuals. As mentioned above, the ADL datasets consist of 35 days of labelled data. This includes 14 days of labelled for "Person-A" and 21 days of labelled data for "Person-B". We use 50% of the total available datasets for learning the models, the remaining 50% is used for testing purpose. The testing datasets are divided into sequence of transitions corresponding to each day and hence we get 7 days in the test dataset for "Person-A" and 10 days in the test dataset for "Person-B". The sequence of transitions corresponding to each test day in test dataset is input to both the models and the amount of increase in code length is noted down in bits. The model that minimally increases the code length is more probable to have generated that sequence.

The increase in code length is quantified in the Figures 18.16 and 18.17 on each HPFSM model on the test dataset belonging to the two individuals for each observed day. Inferred HPFSM (Person-A) model shows a lesser increase in code length for its own test data of every test day, whereas the inferred HPFSM (Person-B) shows a greater amount of increase in code length for "Person-A's" test data of each test day. This results

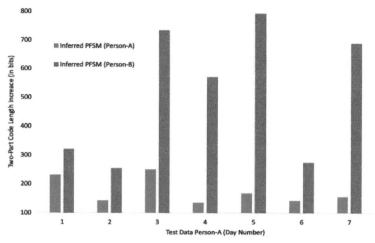

Figure 18.16   Code Length increase in bits in the inferred HPFSM models on test data of Person-A.

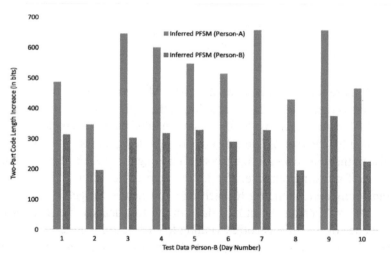

Figure 18.17   Code Length increase in bits in the inferred HPFSM models on test data of Person-B.

to arrive at the conclusion that inferred HPFSM (Person-A) model correctly predicts "Person-A's" movements for each test day as belonging to inferred HPFSM (Person-A) model, then what the other model would do for test day of "Person-A". Similarly inferred HPFSM (Person-B) model correctly predicts the model from the movements, if the movements belong to "Person-B", by showing a lesser increase in two-part code length.

18.6 Summary

This chapter investigated a new learning method based on hierarchical construction of a simple PFSM model called as an HPFSM model. The HPFSM model encoding of both the structure and the data were discussed. The benefits of the construction were understood in viewpoint of the real things happening around us. We discussed the example of a multilingual person and the activities performed by an individual in his daily life if that involves moving around to various places. Most of the time the movements are local and then there is absolutely no need to encode them using one big machine. We can rather construct small machines and combine them somehow so that the purpose is achieved. This is how the construction of hierarchical PFSMs was driven by the need of representing things in a more concise manner and the obvious advantage is a less code length model that still represents the same grammar.

The need for constructing the HPFSM model was then followed by an artificial design of an HPFSM model to understand what the model looks like. Using the same design, we discussed the two-part code length calculation using MML. The first part code length that encodes the model was summarized by Equation (18.7) and the second part code length that encodes the data using the model was summarized by Equation (18.9).

Finally, two different experiments were performed to show the benefits of hierarchical encoding over the non-hierarchical style of encoding. Under the first experimental set up, we considered two artificially created HPFSM models. Different datasets were generated out of the artificially created HPFSM models. The datasets consisted of variable number of random strings from 2 to 5000. In the first artificially created HPFSM model (Example 1 Figure 18.7), the initial probabilities of transition assumed for string generation were considered uniform. Whereas in the second artificially created HPFSM model (Example-2 Figure 18.6, the initial probabilities of transition were considered variable with the internal transitions set to high probabilities and outer transitions set to low probabilities. The two-part encoding cost was then compared with the non-hierarchical PFSM (inferred PFSM) model and the single-state model. The datasets with strings greater than 50 were encoded in a cheaper way by the HPFSM model as compared to the other models, eventually leading to the conclusion that the HPFSM models offered a cheaper way of encoding if the datasets were hierarchical.

The second experiment was performed on the real datasets from the UCI repository. The datasets were the Activities of Daily Living (ADL) datasets. The HPFSM model construction from the datasets was discussed followed by an analysis using the models. The HPFSM model code length, again, in this case also gave a better performance in terms of encoding as compared to the other models. The model learning or encoding was followed by an analysis using the models. We used prediction as the evaluating criteria to test the models. The HPFSM models were correctly able to predict the class from the instances of test data from the ADL datasets.

References

[1] D. Angluin. On the complexity of minimum inference of regular sets. Information and Control, 39(3):337–350, 1978.

[2] P. Cheeseman. On finding the most probable model. In Computational models of scientific discovery and theory formation/edited by Jeff Shrager and Pat Langley. Morgan Kaufmann Publishers, 1990.

[3] C. H. Clelland. Assessment of candidate PFSA models induced from symbol datasets. In Tech Report TR-C95/02. Deakin University, Australia, 1995.

[4] C. H. Clelland and D. A. Newlands. PFSA modelling of behavioural sequences by evolutionary programming. In Proceedings of the Conference on Complex Systems: Mechanism for Adaptation, pages 165–172. Rockhampton, Queensland: IOS Press, 1994.

[5] M. S. Collins and J. J. Oliver. Efficient induction of finite state automata. In Proceedings of the Thirteenth conference on Uncertainty in artificial intelligence, pages 99–107. Morgan Kaufmann Publishers Inc., 1997.

[6] J. Feldman. Some decidability results on grammatical inference and complexity. Information and control, 20(3):244–262, 1972.
B. R. Gaines. Behaviour/Structure Transformation under Uncertainty. International Journal of Man-Machine Studies, 8:337–365, 1976.

[7] M. P. Georgeff and C. S. Wallace. A general selection criterion for inductive inference. In European Conference of Artificial Intelligence (ECAI) 84, 473–482, 1984.

[8] E. M. Gold. Complexity of automaton identification from given data. Information and Control, 37(3):302–320, 1978.

[9] P. Hingston. Inference of regular languages using model simplicity. Australian Computer Science Communications, 23(1):69–76, 2001.

[10] R. Lenhardt. Probabilistic Automata with Parameters. Oriel College University of Oxford, 2009.

[11] J. J. Oliver and D. Hand. Introduction to minimum encoding inference. Technical Report No. 94/205, Department of Computer Science, Monash University, Australia, 1994.

[12] F. J. Ordonez, P. de Toledo, and A. Sanchis. Activity recognition using hybrid generative/discriminative models on home environments using binary sensors. Sensors, 13(5):5460–5477, 2013.

[13] A. Raman and J. Patrick. 14 Linguistic similarity measures using the minimum message length principle. Archaeology and Language, I: Theoretical and Methodological Orientations, 262, 1997a.

[14] A. V. Raman. An information theoretic approach to language relatedness: a dissertation submitted in partial fulfilment of the requirements for the degree of Doctor of Philosophy in Information Systems at Massey University. PhD thesis, Massey University, 1997.

[15] A. V. Raman and J. D. Patrick. Beam search and simba search for PFSA inference. Technical report, Tech Report 2/97, Massey University Information Systems Department, Palmerston North, New Zealand, 1997b.

[16] A. V. Raman, J. D. Patrick, and P. North. The sk-strings method for inferring PFSA. In Proceedings of the workshop on automata induction, grammatical inference and language acquisition at the 14th international conference on machine learning (ICML97), 1997.

[17] K. Rice. Bayesian Statistics (a very brief introducion). Biostat, Epi. 515, 2014.

[18] V. Saikrishna, D. L. Dowe, and S. Ray. MML Inference of Finite State Automata for Probabilistic Spam Detection. In Proceedings of International Conference on Advances in Pattern Recognition (ICAPR), pages 1–6. IEEE Computer Society Press, 2015.

[19] M. Sipser. Introduction to the Theory of Computation, volume 2. Thomson Course Technology Boston, 2006.

[20] C. S. Wallace. Statistical and Inductive Inference by Minimum Message Length. Information Science and Statistics. Springer Science and Business Media, Spring Street, NY, USA, 2005.

Solution to Exercises

Solution (1)
Note the following sequence of weather events

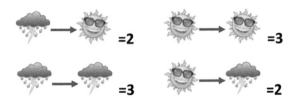

Starting from either sunny or rainy states, the number of instances when the system changed states are shown above. Hence, we can derive the probabilities by normalising the values given above.
Hence:

(a) From the figure there are three instances when a sunny day followed a previous sunny day. This gives the probability $p(s|s) = \frac{3}{5} = 0.6$
(b) There are two days that it rained after a sunny previous day. This gives the probability $p(r|s) = \frac{2}{5} = 0.4$
(c) There are three days that it rained when the previous day was rainy. This also gives the probability $p(r|r) = \frac{3}{5} = 0.6$
(d) There are five rainy days and there are two instances when a rainy day followed a previous rainy day. This gives the probability $p(r|r) = \frac{2}{5} = 0.4$

Solution (3)
$$A = \begin{bmatrix} 0.6 & 0.4 \\ 0.4 & 0.6 \end{bmatrix}$$

Solution (4)

327

The emission probability is computed as a proportion based on the state the system is at. At the sunny state Ireh is happy three times out of four and grumpy once.

Hence the probabilities are:

$$p(h|s) = \frac{3}{4}; \; p(g|s) = \frac{1}{4}$$

$$p(h|r) = \frac{3}{7}; \; p(g|r) = \frac{4}{7}$$

Solution (6)

$$\beta = \begin{bmatrix} \dfrac{3}{4} & \dfrac{1}{4} \\ \dfrac{3}{7} & \dfrac{4}{7} \end{bmatrix}$$

Solution (7)

The answer to this problem depends on the transition probability. Note the question has nothing to do with whether Ireh is happy or grumpy. Hence we can use the transition matrix or the transition diagram to write the following expressions:

$$S = 0.6S + 0.4R \quad (i)$$
$$R = 0.4S + 0.6R \quad (ii)$$

Where S and R refer to the fact that the day is sunny or rainy. We know also that the probability that the day is sunny or rainy must sum to one. Hence the third equation to write is

$$S + R = 1 \quad (iii)$$
$$S = 0.6S + 0.4(0.4S + 0.6R)$$
$$= 0.76S + 0.24R$$

or

$$0.24S = 0.24R$$

Hence S = R and using the equation (iii), S = R = 0.5.

Index

About the Author

Prof Johnson I. Agbinya obtained PhD in microwave radar systems from La Trobe University, Melbourne Australia, MSc (Research) in electronic control from the University of Strathclyde Glasgow Scotland and BSc in Electronic/Electrical Engineering from Obafemi Awolowo University (OAU), Ife Nigeria. He is currently Head, School of Information Technology and Engineering at Melbourne Institute of Technology (MIT), Australia. He was Senior Research Scientist at the Commonwealth Scientific Research Organisation (CSIRO) for nearly a decade before joining Vodafone Australia as Research Manager where he contributed to the design of its 3G network. He subsequently joined the University of Technology as Senior lecturer in the Faculty of Engineering and IT from where he moved back to La Trobe University as Associate Professor. He is currently a full Professor of Engineering at Melbourne Institute of Technology, Australia.

Prof Agbinya's service to humanity is focused on training emerging African Scientists, lecturers and technocrats through a number of African countries including Sudan where he is a Research Professor at the Sudan University of Science and Technology in Khartoum, Nelson Mandela African Institute of Science and Technology Arusha Tanzania as an Adjunct Professor until 2018 and Tshwane University of Technology Pretoria South Africa as Adjunct Professor in ICT with a PhD. He was Professor Extraordinaire in Computer Science at the University of the Western Cape in Cape Town and Extraordinary Professor at the University of Witwatersrand Johannesburg.

Prof Agbinya is a member of Pan African Australasian Diaspora Network (PAADN), member of the Nigerian Society of Engineering and Fellow of African Scientific Institute (ASI).

He has published extensively including ten mobile communication, sensor and data analytics textbooks and over three hundred and fifty journals and conference articles. His technical expertise is in the areas of Mobile Communications, electronic remote sensing, signal processing, wireless power transfer, Internet of Things, biometrics, electrical energy and machine to machine communications and artificial intelligence. He is the founder of the annual conferences IB2COM and African conference called Pan African Conference on Science, Computing and Telecommunication (PACT). He also founded the African Journal of Information and Communication Technology (AJICT). He is currently on the editorial boards of several international journals in ICT and sensors, and editorial consultant at River Publishers Denmark.